Springer Tracts in Natural Philosophy

Volume 17

George R. Gavalas

Nonlinear Differential Equations of Chemically Reacting Systems

With 10 Figures

Springer-Verlag New York Inc. 1968

Professor George R. Gavalas, Chemical Engineering, California Institute of Technology. 1201 East California Boulevard, Pasadena, Cal. 91109/USA

ISBN 978-3-642-87645-5 ISBN 978-3-642-87643-1 (eBook)
DOI 10.107/978-3-642-87643-1

 Library of Congress Catalog Card Number 68-31619. Title-No. 6745
Softcover reprint of the hardcover 1st edition 1968

To my father, Lazaros

Preface

In recent years considerable interest has developed in the mathematical analysis of chemically reacting systems both in the absence and in the presence of diffusion. Earlier work has been limited to simple problems amenable to closed form solutions, but now the computer permits the numerical solution of complex systems of nonlinear differential equations. The numerical approach provides quantitative information, but for practical reasons it must be limited to a rather narrow range of the parameters of the problem. Consequently, it is desirable to obtain broader qualitative information about the solutions by investigating from a more fundamental mathematical point of view the *structure* of the differential equations. This theoretical approach can actually complement and guide the computational approach by narrowing down trial and error procedures, pinpointing singularities and suggesting methods for handling them. The study of the structure of the differential equations may also clarify some physical principles and suggest new experiments. A serious limitation of the theoretical approach is that many of the results obtained, such as the sufficient conditions for the stability of the steady state, turn out to be very conservative. Thus the theoretical and computational approaches are best used together for the purpose of understanding, designing, and controlling chemically reacting systems.

The present monograph is intended as a contribution to the theory of the differential equations describing chemically reacting systems. The main topics treated are the a priori bounds and the existence of solutions, the conditions for uniqueness and stability of solutions and the asymptotic behavior of solutions. Most of these problems have been treated by using methods from nonlinear functional analysis and especially fixed point methods. The closely related theory of bifurcation points seems to hold promise for further progress.

It is hoped that the monograph will be of interest to both the theoretical engineer and the applied mathematician. The former will notice that chemical reactions have been considered in the framework of three specific systems, the uniform isolated system (batch reactor), a uniform open system (stirred-tank reactor), and a distributed system (catalyst pellet). However, most of the methods presented should be useful in analyzing other systems such as fixed bed reactors. The complex systems of biological reactions seem also to offer an exciting ground for mathematical analysis. The applied mathematician will hopefully find an element of elegance and generality in the differential equations of chemical systems. These equations seem to be an ideal subject for the application and further development of nonlinear analysis.

I have tried to cite references with the latest information on each subject discussed and have not attempted to reference the earliest contributors. Consequently, I have emphasized recent papers and textbooks which in turn give access to the earlier work.

I am indebted to Professor RUTHERFORD ARIS of the University of Minnesota for encouraging me to undertake this work and for his critical review of the manuscript. I am also indebted to Professor DAN LUSS of the University of Houston for many constructive criticisms and comments. I would like to thank Miss EVELYN ANDERSON and Mrs. RUTH BARNETT for typing the manuscript and Mr. FRANK CHU for assisting with the preparation of the manuscript.

Pasadena, August 1968 GEORGE R. GAVALAS

Table of Contents

Chapter 1

Uniform Systems with Chemical Change

1.1. Stoichiometry and Kinetics of Chemical Reactions

A homogeneous system can be locally characterized by the velocity, the concentrations of all the chemical species, and one thermodynamic variable, such as the temperature or the internal energy. These variables will be called the state variables. A nonuniform or "distributed" system is completely specified by the values of the state variables at each point of the system, while a uniform system, not having any space variations, is specified by the values of the state variables at a single point.

The evolution of systems with chemical change is described by the equations of conservation of momentum, energy, and mass of each chemical species. The conservation equations for the chemical species contain source terms due to the chemical reactions. Certain linear combinations of these source terms vanish due to the conservation of atomic species in each chemical reaction. This property has very important implications for the mathematical structure of the differential equations describing uniform and distributed reacting systems.

Stoichiometry is a term indicating the algebraic relations applying to composition changes with no regard to the rates of these changes. An elementary account of stoichiometry may be found in ARIS [3, 8]. In the last few years significant advances have been made in the algebra of chemical reactions. We should mention the work of ARIS [4, 5], HORIUTI [19], BOWEN [10], SELLERS [37, 38]. Some of this work is of a very general and abstract nature. In this section we shall discuss only a few elementary stoichiometric properties which will be necessary for our study of the differential equations of reacting systems.

Let us consider a homogeneous mixture of N chemical species denoted by the symbols: $\mathscr{M}_1, \ldots, \mathscr{M}_N$. These species consist of a number of atomic species: $\mathscr{A}_1, \ldots, \mathscr{A}_M$, and β_{ij} will denote the number of atoms $\mathscr{A}_j$ in the species $\mathscr{M}_i$.

A chemical reaction is symbolized by

$$\sum_{i=1}^{N} \nu_i \mathscr{M}_i = 0 \tag{1.1.1}$$

and signifies a possible interconversion of the participating species in amounts δn_i proportional to the stoichiometric coefficients ν_i:

$$\frac{\delta n_i}{\delta n_k} = \frac{\nu_i}{\nu_k}. \tag{1.1.2}$$

By convention, the species $\mathcal{M}_i$ is called a reactant or a product of the reaction according to whether $\nu_i > 0$ or $\nu_i < 0$.

For a number of R' simultaneous reactions we shall use the notation:

$$\sum_{i=1}^{N} \nu_{ij} \mathcal{M}_i = 0 \qquad j = 1, \ldots, R' \tag{1.1.3}$$

where ν_{ij} is the stoichiometric coefficient of the species $\mathcal{M}_i$ in the jth reaction. The matrix $\boldsymbol{\nu} = (\nu_{ij})$ is called the stoichiometric matrix.

The reaction system (1.1.3) is capable of describing the chemical change if with each reaction we can associate a function f_j of the $N+1$ state variables (e.g. the N concentrations and the temperature) such that:

$$\left(\frac{dn_i}{dt}\right)_{\text{ch}} = \sum_{j=1}^{R'} \nu_{ij} f_j(c_1, \ldots, c_N, T) = F_i(c_1, \ldots, c_N, T) \tag{1.1.4}$$

is the total production of $\mathcal{M}_i$ in moles per unit volume per unit time due to chemical reactions. The evolution of a homogeneous and uniform system of constant volume is described by the species conservation equations

$$\frac{dc_i}{dt} = \sum_{j=1}^{R'} \nu_{ij} f_j(c_1, \ldots, c_N, T) = F_i(c_1, \ldots, c_N, T) \tag{1.1.5}$$

and the energy conservation equation. In distributed systems the term in the right of Eq. (1.1.5) appears as a source term in the species conservation equations. The functions $f_1, \ldots, f_{R'}$; $F_1, \ldots, F_N$ are called the *rate functions*, or the *rate laws*, or the *kinetics*, or simply the *reaction rates*.

Often, there are more than one equivalent sets of reactions capable of describing a given chemical change, each set having its own rates f_j. The rates F_i, however, are the same for all equivalent sets. The descriptions in terms of $f_1, \ldots, f_{R'}$ and $F_1, \ldots, F_N$ are both useful.

It is often desirable to describe composition changes by the smallest possible number of reactions. This leads to the concept of independent reactions.

Definition 1.1.1. The reactions of Eq. (1.1.3) are called *independent* if the rank of the stoichiometric matrix $\boldsymbol{\nu}$ is equal to the number of the reactions.

If the rank of $\boldsymbol{v}$ is $R<R'$ then there exist R linearly independent rows which, by renumbering if necessary, can be considered as the first R rows. The remaining $R'-R$ rows can be expressed as combinations of the first R:

$$v_{ij}=\sum_{l=1}^{R}\gamma'_{jl}\,v_{il} \qquad j=R+1,\dots,R';\ i=1,\dots,N. \tag{1.1.6}$$

By substituting Eq. (1.1.6) in Eq. (1.1.4) there is obtained

$$F_i=\sum_{j=1}^{R}v_{ij}\,g_j \tag{1.1.7}$$

where

$$g_j=f_j+\sum_{l=R+1}^{R'}\gamma'_{lj}\,f_l. \tag{1.1.8}$$

It follows that chemical change can always be described by a system of independent reactions.

Whether or not a system of reactions is capable of describing composition changes is a matter of experimental investigation. The number of independent reactions, however, has an upper bound, independently of experimental information. This upper bound results from the conservation of atomic species, i.e.

$$\sum_{i=1}^{N}v_{ij}\,\beta_{il}=0 \qquad j=1,\dots,R';\ l=1,\dots,M. \tag{1.1.9}$$

If the rank of the matrix $\boldsymbol{\beta}$ is R_β, Eq. (1.1.9) includes $R_\beta\geqq 1$ linearly independent relations among the columns of the matrix $\boldsymbol{v}$ so that

$$R\leqq N-R_\beta. \tag{1.1.10}$$

Let us now examine briefly how the concepts of stoichiometry and kinetics relate to experiment. The quantities that can be directly measured in the laboratory are the pressure, temperature, concentrations and other functions of the state variables such as the thermal and electrical conductivity. From such measurements it is possible to determine the number of independent reactions, but no distinction can be made among equivalent sets of reactions.

The determination of the rates $f_1,\dots,f_R$ or $F_1,\dots,F_N$ from experimental data is a task both difficult and of limited accuracy. The usual practice is to construct a kinetic model leading to rate functions with a few parameters which are then chosen to give the best fit, in a statistical sense, to the data. The kinetic model is "good" if it describes well a large amount of data. There exists considerable literature on the use of statistical techniques for the determination of kinetic parameters, e.g. [22, 28, 12]. For industrial work, it is sufficient to have simple rate functions, which may have to be largely empirical. These ought to agree reasonably well with measurements over the conditions of the particular application.

Chemical Kinetics, as a branch of chemistry, is concerned with the determination of the rates and the *mechanism* of reactions. The concept of the reaction mechanism is based on the concept of an *elementary reaction*. This is a reaction that corresponds, in a sense, to a single molecular collision. A reaction is generally composed of a number of elementary reactions which constitute the reaction mechanism.

For example, the reaction

$$H_2 + Br_2 \xrightarrow{f} 2\,HBr \tag{1.1.11}$$

cannot be explained by a single molecular encounter. It can be explained, however, in terms of a number of elementary reactions such as

$$Br_2 + M \underset{f_2}{\overset{f_1}{\rightleftarrows}} 2\,Br + M, \tag{1.1.12}$$

$$Br + H_2 \underset{f_4}{\overset{f_3}{\rightleftarrows}} HBr + H, \tag{1.1.13}$$

$$H + Br_2 \xrightarrow{f_5} HBr + Br \tag{1.1.14}$$

where M is H_2 or Br_2.

If c_1, c_2, c_3, c_4, c_5 are the concentrations of the species H_2, Br_2, HBr, H, Br the rates of the above five elementary reactions may be represented by:

$$f_1 = k_1\, c_2 (c_1 + c_2), \tag{1.1.15}$$

$$f_2 = k_2\, c_5^2 (c_1 + c_2), \tag{1.1.16}$$

$$f_3 = k_3\, c_5\, c_1, \tag{1.1.17}$$

$$f_4 = k_4\, c_3\, c_4, \tag{1.1.18}$$

$$f_5 = k_5\, c_4\, c_2 \tag{1.1.19}$$

where the rate constants k_i depend on the temperature only:

$$k_i = k_i^*\, e^{-\frac{E_i}{R_g T}}. \tag{1.1.20}$$

The five elementary reactions (1.1.12)–(1.1.14) constitute a mechanism for the reaction (1.1.11). The rate of formation of HBr can be obtained by combining the rate expressions (1.1.15)–(1.1.19) and would involve the concentrations of the free atoms H, Br which may be present in very small quantities. For a broad range of conditions, it is possible to eliminate the concentrations of H, Br from the overall rate by setting

$$F_4 = f_3 - f_4 - f_5 = 0, \tag{1.1.21}$$

$$F_5 = 2f_1 - 2f_2 - f_3 + f_4 + f_5 = 0. \tag{1.1.22}$$

These equations give

$$c_4 = \frac{k_3 \left(\frac{k_1}{k_2}\right)^{\frac{1}{2}} c_1 c_2^{\frac{1}{2}}}{k_5 c_2 + k_4 c_3}, \qquad c_5 = \left(\frac{k_1}{k_2}\right)^{\frac{1}{2}} c_2^{\frac{1}{2}}, \tag{1.1.23}$$

$$f = f_4 - f_3 = \frac{k_3 \left(\frac{k_1}{k_2}\right)^{\frac{1}{2}} c_1 c_2^{\frac{1}{2}}}{1 + \frac{k_4 c_3}{k_5 c_2}}. \tag{1.1.24}$$

This is the *steady state approximation* whose mathematical justification has been considered in [9, 18].

Two theories have been developed for the prediction of rates of elementary reactions. One is the collision theory [15] based on the classical kinetic theory of gases and applicable to gaseous reactions only. The other is the transition state theory [27] making use of statistical mechanics and quantum mechanics. As an example, the collision theory predicts the rate of the elementary reaction (1.1.14) as

$$f_5 = k\, c_2\, c_4\, T^{\frac{1}{2}}\, e^{-\frac{E}{R_g T}}. \tag{1.1.25}$$

The activation energy E can in principle be calculated from quantum mechanics. This calculation is so complex, however, that it has been performed only for the simplest molecules. The rate constant k can be calculated from the collision and the transition state theory, but very often the calculated and experimental values deviate considerably. Although both theories involve approximations and physically questionable assumptions, expressions such as (1.1.25) seem to correlate well experimental data and are, consequently, very useful.

We should finally mention recent theoretical and experimental studies of chemical reactions on a more detailed microscopic level. In these studies the collision dynamics are studied in detail by classical or quantum mechanics in order to obtain the *cross section* for the reaction. The cross section is a function of relative speed, orientation, and possibly the internal state of the colliding molecules and can be averaged over these variables to give the macroscopic reaction rate. This very intriguing theoretical approach is unfortunately extremely complicated and so far has been applied only in the simplest reactions. References on this work may be found in [1].

We may conclude that reaction rates can be entirely phenomenological or can be derived from microscopic theories of varying level of detail. The parameters in the rates have to be obtained almost always from experiment. Nevertheless, the theory is very valuable because it provides the functional form of the rates and estimates for the parameters.

In this work attention will be focused on nonlinear differential equations arising from nonlinear reaction kinetics. For the special, but important case of linear reaction kinetics, we refer the reader to the comprehensive treatment of WEI and PRATER [43].

1.2. Invariant Manifolds and Extents of Reactions

The present section is concerned with the following basic property of reacting systems: The state variables of a closed uniform system remain at all times in a bounded R-dimensional subset of the $N+1$ dimensional state space. To derive this property, we shall first discuss the way in which conservation and thermodynamic principles restrict the admissible class of reaction rate functions.

Any chemical kinetic model should in principle be consistent with the conservation of atomic species, the nonnegativity of concentrations and temperature, and the first and second laws of thermodynamics. The first requirement was considered in the previous section and led to Eq. (1.1.9), for the elements of the stoichiometric matrix. The nonnegativity of concentrations and temperature will be insured by Postulate 1.2.1, formulated below. The first law of thermodynamics is included in the energy equation describing a reacting system. These three requirements will be assumed to be satisfied throughout this work.

The requirement that the kinetic model obeys the second law of thermodynamics is often inconvenient as it involves expressions for entropy, chemical potentials, etc., which are not available for all systems. Moreover, some empirical kinetic models, which describe chemical systems approximately, may not be consistent with the second law over the entire range of interest. Yet these models, often the only ones available, are quite useful. We shall study, therefore, kinetic models which satisfy the other requirements mentioned above but do not necessarily obey the second law of thermodynamics. Section 1.5 is concerned with kinetic models satisfying the second law of thermodynamics. For a somewhat different formulation of the admissible class of reaction kinetic models see the paper by WEI [40].

The evolution, in time, of a uniform reacting system depends not only on its internal dynamics, but also on its interaction with its surroundings. This interaction in many cases takes the form of a simple thermodynamic constraint imposed on the system, for example constant temperature and pressure, or isolation. In the following few sections we shall consider a system of constant volume which does not exchange mass or energy with its surroundings, i.e., an *isolated* system. Other *closed* systems, i.e., those which exchange energy but no mass with their surroundings can be treated in a similar fashion. Section 1.7 treats one important type of *open* uniform systems. The notation will be as follows: $\mathbf{u}=(c_1, \ldots, c_N, T)$ is the state vector, E_{N+1} is the corresponding $N+1$ dimensional Euclidean state space and E_{N+1}^+ is its positive orthant:

$$E_{N+1}^+ = \{\mathbf{u}\colon c_1, \ldots, c_N, T \geqq 0\}. \tag{1.2.1}$$

When only the concentrations are of interest we shall let $\mathbf{c}=(c_1, \ldots, c_N)$ and denote by, E_N, E_N^+ the corresponding Euclidean space and its positive orthant.

It is convenient to consider an isolated system of unit volume whereby U, the total internal energy of the system, is also the internal energy per unit volume. The time evolution of this system is described by the equations:

$$\frac{dc_i}{dt} = \sum_{j=1}^{R} \nu_{ij} f_j \qquad i=1, \ldots, N, \tag{1.2.2}$$

$$U(c_1, \ldots, c_N, T) \equiv U_0 \tag{1.2.3}$$

where U_0 is a constant. The admissible class of functions $f_1, \ldots, f_R$, U is defined by

Postulate 1.2.1. The functions $f_1, \ldots, f_R$, U are defined and continuous in E_{N+1}. For any $c_i=0$

$$\nu_{ij} f_j \geqq 0 \qquad j=1, \ldots, R. \tag{1.2.4}$$

For $T=0$

$$f_j = 0 \qquad j=1, \ldots, R. \tag{1.2.5}$$

For $c_1 \geqq 0, \ldots, c_N \geqq 0$, $c_1 + \cdots + c_N > 0$

$$\lim_{T \to \infty} U(c_1, \ldots, c_N, T) = +\infty. \tag{1.2.6}$$

For $T_1 < T_2$

$$U(c_1, \ldots, c_N, T_1) < U(c_1, \ldots, c_N, T_2). \tag{1.2.7}$$

A few comments on this postulate are in order. First, the functions $f_1, \ldots, f_R$, U have physical meaning only in a bounded subregion of E_{N+1}^+. Being defined in such a subregion they can always be extended continuously over E_{N+1}^+. The actual extension, however, is immaterial because the trajectories emanating from the physically meaningful subregion will remain in this subregion. The definition of $f_1, \ldots, f_R, U$ in the whole space E_{N+1} is a matter of mathematical convenience and does not impair the physical value of the model. Conditions (1.2.4), (1.2.5) are physically obvious and insure the nonnegativity of concentrations and temperature. More precisely, a trajectory starting in E_{N+1}^+ will remain in this region at all subsequent times. Conditions (1.2.6), (1.2.7) are also physically obvious and insure that Eq. (1.2.3) can be solved uniquely for T in terms of $c_1, \ldots, c_N$. We shall denote this solution by $T=W(c_1, \ldots, c_N, U_0)$, where W is a continuous function such that

$$U(c_1, \ldots, c_N, W(c_1, \ldots, c_N, U_0)) \equiv U_0. \tag{1.2.8}$$

It will be shown now that trajectories remain in a bounded R-dimensional region of E^+_{N+1}. Without loss in generality it can be assumed that the R chemical reactions are independent and since $R<N$ there exist $N-R$ independent linear relations among the columns of the stoichiometric matrix $\boldsymbol{\nu}$:

$$\sum_{i=1}^{N} \gamma_{il}\,\nu_{ij}=0 \qquad j=1,\ldots,R;\; l=1,\ldots,N-R. \tag{1.2.9}$$

R_β of these relations can be chosen from Eq.(1.1.9):

$$\gamma_{il}=\beta_{il}\geqq 0 \qquad i=1,\ldots,N;\; l=1,\ldots,R_\beta. \tag{1.2.10}$$

Furthermore, each chemical species contains at least one atomic species therefore

$$\sum_{l=1}^{M} \beta_{il}>0 \qquad i=1,\ldots,N. \tag{1.2.11}$$

Multiplication of Eq.(1.2.1) by γ_{il}, summation over i and integration gives

$$\frac{d}{dt}\sum_{i=1}^{N} \gamma_{il}\,c_i=0 \qquad l=1,\ldots,N-R \tag{1.2.12}$$

or

$$\sum_{i=1}^{N} \gamma_{il}(c_i-c_{i0})=0 \qquad l=1,\ldots,N-R \tag{1.2.13}$$

which for $l=1,\ldots,R_\beta$ simply express the conservation of the atomic species $\mathscr{A}_l$. The conservation of internal energy is described by

$$U(c_1,\ldots,c_N,T)=U(c_{10},\ldots,c_{N0},T_0)=U_0 \tag{1.2.14}$$

where $\mathbf{u}_0=(c_{10},\ldots,c_{N0},T_0)$ is a fixed point in the trajectory, for example the initial point. Eq.(1.2.13), (1.2.14) define an R-dimensional manifold whose intersection with E^+_{N+1} will be denoted by $\Gamma(u_0)$, indicating that the manifold includes the point $\mathbf{u}_0$.

The following remarks are obvious from the definition of $\Gamma(u_0)$. Two points $\mathbf{u}, \mathbf{u}_0$ belong to the same manifold if and only if they lie in E^+_{N+1} and satisfy Eqs.(1.2.13), (1.2.14). A trajectory of Eqs.(1.2.1), (1.2.2) having a point in $\Gamma(u_0)$ lies entirely in this manifold. It is clear then that E^+_{N+1} is partitioned by Eqs.(1.2.13), (1.2.14) in an infinite number of mutually exclusive manifolds, invariant under the differential Eqs.(1.2.1), (1.2.2).

Definition 1.2.1. The set of points in E^+_{N+1} satisfying Eqs.(1.2.13), (1.2.14) will be denoted by $\Gamma(u_0)$ and called the *invariant manifold* corresponding to the point $\mathbf{u}_0$. In the theory of differential equations a set such as $\Gamma(u_0)$ is sometimes called an integral manifold.

The following simple theorem will be used in subsequent sections.

Theorem 1.2.1. An invariant manifold $\Gamma(u_0)$ is a closed, connected, and bounded set.

Proof. The closedness follows directly from Eqs. (1.2.13), (1.2.14). The connectedness can be proved by showing that $\Gamma(u_0)$ is arcwise connected. Let $\mathbf{u}, \bar{\mathbf{u}}$ be any two points in $\Gamma(u_0)$, then the arc

$$c_i(s) = s c_i + (1-s)\bar{c}_i \qquad i=1,\ldots,N, \tag{1.2.15}$$

$$T(s) = W(c_1(s),\ldots,c_N(s), U_0) \tag{1.2.16}$$

lies in $\Gamma(u_0)$ for all $s \in [0,1]$. The boundedness of the variables $c_1,\ldots,c_N$ follows from Eqs. (1.2.10), (1.2.11) in combination with Eq. (1.2.13). Finally, the boundedness of $c_1,\ldots,c_N$ and the continuity of the function $W(c_1,\ldots,c_N, U_0)$ imply that T is also bounded.

Physically, the boundedness of $\Gamma(u_0)$ is due to the conservation of atomic species and energy and the nonnegativity of the variables $c_1,\ldots,c_N, T$. Let us use the notation

$$c_{im} = \min_{\Gamma(u_0)} c_i, \qquad c_{iM} = \max_{\Gamma(u_0)} c_i, \tag{1.2.17}$$

$$T_m = \min_{\Gamma(u_0)} T, \qquad T_M = \max_{\Gamma(u_0)} T. \tag{1.2.18}$$

A trajectory $\mathbf{u}(t)$ of Eqs. (1.2.2), (1.2.3) passing from $\mathbf{u}_0$ lies entirely in $\Gamma(u_0)$, therefore for any t

$$c_{im} \leqq c_i(t) \leqq c_{iM}, \qquad T_m \leqq T(t) \leqq T_M. \tag{1.2.19}$$

The bounds c_{im}, c_{iM}, T_m, T_M depend on the initial conditions but are independent of the reactions rates. Moreover, these bounds are obtained without a knowledge of the solution of Eqs. (1.2.2), (1.2.3), hence they are called *a priori bounds.* The a priori bounds are important in the mathematical study of Eqs. (1.2.2), (1.2.3) and are also useful in practical calculations.

Since any given trajectory of Eqs. (1.2.2), (1.2.3) lies entirely in an R-dimensional manifold $\Gamma(u_0)$, it is often convenient to introduce in $\Gamma(u_0)$ intrinsic coordinates $\xi_1,\ldots,\xi_R$ by means of the equations

$$c_i - c_{i0} = \sum_{j=1}^{R} \nu_{ij}\xi_j \qquad i=1,\ldots,N. \tag{1.2.20}$$

Eq. (1.2.20) and (1.2.3) define a one-to-one mapping between $\Gamma(u_0)$ and its image $\tilde{\Gamma}(u_0)$ in the ξ-space. This mapping may be denoted by

$$\mathbf{u} = \mathbf{u}(\xi; \mathbf{u}_0), \qquad \xi \in \tilde{\Gamma}(u_0), \tag{1.2.21}$$

$$\xi = \xi(\mathbf{u}; \mathbf{u}_0), \qquad \mathbf{u} \in \Gamma(u_0), \tag{1.2.22}$$

$$\tilde{\Gamma}(u_0) = \{\xi : \xi = \xi(\mathbf{u}; \mathbf{u}_0),\ \mathbf{u} \in \Gamma(u_0)\}. \tag{1.2.23}$$

The variable ξ_j represents the contribution of the jth reaction in the change from the state $\mathbf{u}_0$ to the state $\mathbf{u}$ and is called *the extent of the jth reaction.* The extents may also be interpreted as degrees of freedom in the thermodynamic sense.

In terms of the extents, the system of $N+1$ Eqs. (1.2.22), (1.2.23), reduces to a system of R equations

$$\frac{d\xi_j}{dt}=\tilde{f}_j(\xi_1,\ldots,\xi_R) \qquad j=1,\ldots,R \tag{1.2.24}$$

where

$$\tilde{f}_j(\xi_1,\ldots,\xi_R)=f_j(\mathbf{u}(\boldsymbol{\xi};\mathbf{u}_0)). \tag{1.2.25}$$

Example 1.2.1. To illustrate the concepts of this section, we shall consider the two reactions

$$2NO_2-N_2O_4=0, \tag{1.2.26}$$

$$2NO+O_2-2NO_2=0 \tag{1.2.27}$$

and attach subscripts 1, 2, 3, 4 to the species O_2, NO, NO_2, N_2O_4. The internal energy is taken as

$$U(c_1,\ldots,c_N,T)=\sum_{i=1}^{4}c_i\,c_{v\,i}\,T \tag{1.2.28}$$

with constant molar heat capacities $c_{v\,i}$. The manifold passing from the point $\mathbf{u}_0=(0,0,0,c_{40},T_0)$ is given by the equations

$$2c_1+c_2+2c_3+4c_4=4c_4, \tag{1.2.29}$$

$$c_2+c_3+2c_4=2c_{40}, \tag{1.2.30}$$

$$(c_{v1}\,c_1+c_{v\,2}\,c_2+c_{v\,3}\,c_3+c_{v\,4}\,c_4)\,T=c_{v\,4}\,c_{40}\,T_0 \tag{1.2.31}$$

which express the conservation of the atomic species O, N and the internal energy. This is a two dimensional manifold in E_5.

Extents of reaction are introduced by setting

$$c_1=\xi_2, \tag{1.2.32}$$

$$c_2=2\xi_2, \tag{1.2.33}$$

$$c_3=2\xi_1-2\xi_2, \tag{1.2.34}$$

$$c_4-c_{40}=-\xi_1, \tag{1.2.35}$$

$$T=\frac{c_{v\,4}\,c_{40}}{c_{v\,4}\,c_{40}+(2c_{v\,3}-c_{v\,4})\,\xi_1+(2c_{v\,2}+c_{v1}-2c_{v\,3})\,\xi_2}. \tag{1.2.36}$$

Eqs. (1.2.32)–(1.2.36) can be solved uniquely for $\boldsymbol{\xi}$ in terms of $\mathbf{u}$, provided $\mathbf{u}\in\Gamma(u_0)$. Fig. 1.2.1 shows the invariant manifold $\tilde{\Gamma}(u_0)$ which is determined by the inequalities $c_i\geqq 0$ $(i=1,2,3)$ and $T>0$ which can be written in terms of the $\boldsymbol{\xi}$ variables as

$$\xi_2\geqq 0, \quad \xi_1-\xi_2\geqq 0, \quad \xi_1\leqq c_{40}, \tag{1.2.37}$$

$$(2c_{v\,3}-c_{v\,4})\,\xi_1+(2c_{v\,2}+c_{v1}-2c_{v\,3})\,\xi_2+c_{v\,4}\,c_{40}\geqq 0. \tag{1.2.38}$$

The boundary of $\tilde{\Gamma}(u_0)$ consists of the points at which one or more of the equality signs holds.

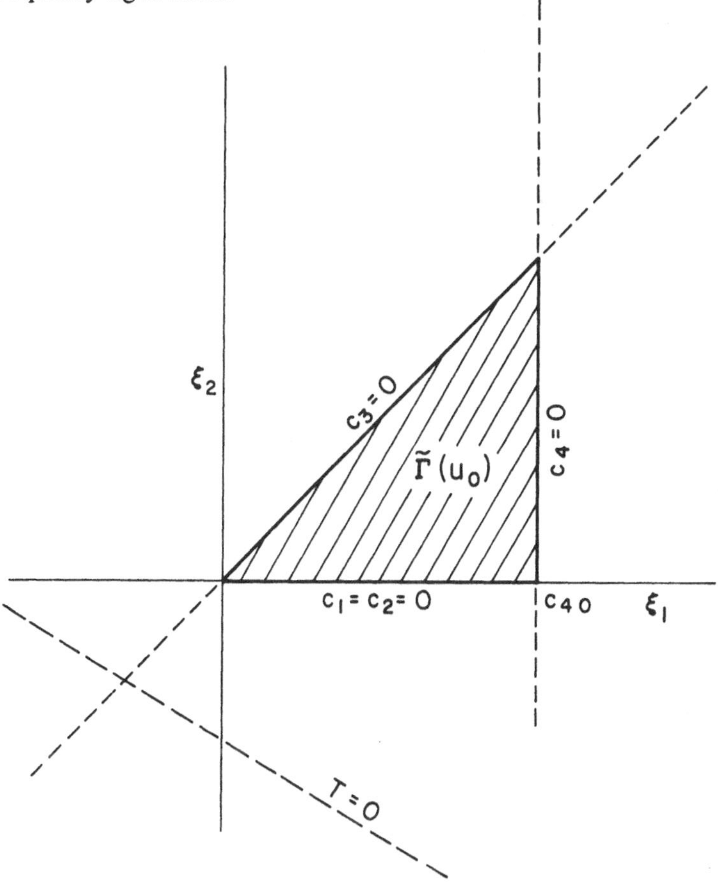

Fig. 1.2.1. An invariant manifold in Example 1.2.1

1.3. Existence and Uniqueness of Solutions

In the present section we shall show that the initial value problem

$$\frac{d\xi_j}{dt} = \tilde{f}_j(\xi_1, \ldots, \xi_R) \qquad j = 1, \ldots, R, \tag{1.3.1}$$

$$\xi_j(t_0) = 0 \tag{1.3.2}$$

describing the time evolution of an isolated system, has a solution for all times $t > t_0$ (existence in the large). The existence will be proved by a fixed point method which is ideally suited for the qualitative study of differential equations containing nonlinear source terms due to chemical reactions. Fixed point methods are readily applicable to this type of differential equations due to the existence of bounded invariant manifolds. While the conditions of Postulate 1.2.1 suffice for providing that a solution exists, the solution need not be unique unless an additional condition such as a Lipschitz condition is imposed.

The basic definitions and results from fixed point theory used in this work have been taken from the excellent treatise of KRASNOSEL'SKII [23] and are summarized for convenience in an appendix. In order to use the fixed point method, Eqs. (1.3.1), (1.3.2) are written in the integral form

$$\boldsymbol{\xi}(t)=\int_{t_0}^{t}\tilde{\mathbf{f}}(\boldsymbol{\xi}(\tau))\,d\tau=\mathbf{H}\boldsymbol{\xi} \tag{1.3.3}$$

which defines the nonlinear operator $\mathbf{H}$. Let us consider the Banach space $\mathscr{B}_c$ of all vector valued functions

$$\boldsymbol{\xi}=(\xi_1(t),\ldots,\xi_R(t)) \tag{1.3.4}$$

continuous in the interval $[t_0, t_1]$ where $t_1>t_0$ is arbitrary but fixed. When $\boldsymbol{\xi}$ is evaluated at a given time, it represents an ordinary vector in E_R. The norm in $\mathscr{B}_c$ is defined by

$$\|\boldsymbol{\xi}\|=\sum_{j=1}^{R}\max_{t_0\leqq t\leqq t_1}|\xi_j(t)| \tag{1.3.5}$$

and the Euclidean norm in E_R is denoted by $|\boldsymbol{\xi}|$.

All possible solutions $\boldsymbol{\xi}$ of Eq. (1.3.3), considered as elements of $\mathscr{B}_c$, lie in the closed bounded region $V(u_0)\subset\mathscr{B}_c$ defined by

$$V(u_0)=\{\boldsymbol{\xi}\colon \boldsymbol{\xi}(\tau)\in\tilde{\Gamma}(u_0),\ t_0\leqq\tau\leqq t_1\}. \tag{1.3.6}$$

As with $\tilde{\Gamma}(u_0)$, $V(u_0)$ is bounded independently of the reaction rates. Any $\boldsymbol{\xi}\in V(u_0)$ therefore satisfies

$$\|\boldsymbol{\xi}\|<b \tag{1.3.7}$$

where b depends on the initial conditions $\mathbf{u}_0$ but not on the reaction rates.

The boundary $\partial V(u_0)$ of the region $V(u_0)$ consists of all points $\boldsymbol{\xi}$ such that $\boldsymbol{\xi}(\tau)$ lies on the boundary of $\tilde{\Gamma}(u_0)$ for one or more values of τ. We intend to prove the existence of a solution of Eq. (1.3.3) by showing that the vector field $\mathscr{H}=\mathbf{I}-\mathbf{H}$ has rotation $+1$ on $\partial V(u_0)$. It may happen, however, that the solution lies on the boundary itself, in which case the rotation is not defined. To avoid this difficulty we enclose $V(u_0)$ in the interior of the ball

$$V_1(u_0)=\{\boldsymbol{\xi}\colon \|\boldsymbol{\xi}\|\leqq b\} \tag{1.3.8}$$

where b has been defined by Eq. (1.3.7). To calculate the rotation of $\mathscr{H}$ on $\partial V_1(u_0)$ we first establish a lemma.

Lemma 1.3.1. The operator $\mathbf{H}$ of Eq. (1.3.3) is completely continuous in the region $V_1(u_0)$.

Proof. The following bounds can be obtained for any sequence $\{\boldsymbol{\xi}^{(n)}\} \in V_1(u_0)$

$$\left| \int_{t_0}^{t} \tilde{f}_j(\boldsymbol{\xi}^{(n)}(\tau)) \, d\tau \right| \leqq (t - t_0) f_{jM}, \tag{1.3.9}$$

$$\left| \int_{t_0}^{t} \tilde{f}_j(\boldsymbol{\xi}^{(n)}(\tau)) \, d\tau - \int_{t_0}^{t'} \tilde{f}_j(\boldsymbol{\xi}^{(n)}(\tau)) \, d\tau \right| \leqq |t - t'| f_{jM} \tag{1.3.10}$$

where

$$f_{jM} = \max_{\|\boldsymbol{\xi}\| \leqq b} |\tilde{f}_j(\boldsymbol{\xi})|. \tag{1.3.11}$$

Eqs. (1.3.9), (1.3.10) show that the sequence $\{\mathbf{H}\,\boldsymbol{\xi}^{(n)}\}$ is equicontinuous and uniformly bounded. By ASCOLI's theorem, which holds for both scalar and vector valued functions, there exists a uniformly convergent subsequence of $\{\mathbf{H}\,\boldsymbol{\xi}^{(n)}\}$ which is also convergent in the norm of Eq. (1.3.5). This shows the complete continuity of the operator $\mathbf{H}$.

Theorem 1.3.4. The initial value problem, Eqs. (1.3.1), (1.3.2), has a solution for all $t > t_0$.

Proof. We shall show that the vector fields $\mathscr{H} = \mathbf{I} - \mathbf{H}$ and $\mathbf{I}$ are homotopic on the boundary $\partial V_1(u_0)$. The vector field

$$\mathscr{H}(s) = \mathbf{I} - s\,\mathbf{H} \qquad 0 \leqq s \leqq 1 \tag{1.3.12}$$

is completely continuous and satisfies the conditions (i) and (ii) of definition (A.4). It remains to show that the vector field $\mathscr{H}(s)$ has no null points on $\partial V_1(u_0)$ for any value of s in $[0, 1]$. If $\boldsymbol{\xi}$ is a null point of $\mathscr{H}(s)$

$$\boldsymbol{\xi}(t) = s \int_{t_0}^{t} \tilde{\mathbf{f}}(\boldsymbol{\xi}(\tau)) \, d\tau. \tag{1.3.13}$$

But this is the same problem as Eq. (1.3.3) except that all the rates are multiplied by the factor s. Consequently, $\boldsymbol{\xi} \in V(u_0)$, $\boldsymbol{\xi} \notin \partial V_1(u_0)$. The two vector fields $\mathscr{H}$, $\mathbf{I}$ as homotopic, have the same rotation, $+1$. Hence, $\mathbf{H}$ has at least one fixed point in the interior of $V_1(u_0)$. According to our earlier discussion, this fixed point lies in $V(u_0)$. Finally, since t_1 is arbitrary, the solution exists for all times.

The problem of uniqueness of solutions for differential equations describing systems with chemical change has not been adequately investigated. Uniqueness is expected on physical grounds but is not ensured by the conditions of Postulate 1.2.1 alone as is shown by the following artificial example of a single isothermal reaction.

$$\mathcal{M}_2 - \mathcal{M}_1 = 0, \tag{1.3.14}$$

$$f = \begin{cases} k_1(2a - c_1)^{\frac{1}{2}}, & a \leqq c_1 \leqq 2a \\ k_1 a^{-\frac{1}{2}} c_1, & 0 \leqq c_1 \leqq a. \end{cases} \tag{1.3.15}$$

The initial value problem

$$\frac{dc_1}{dt} = f(c_1), \tag{1.3.16}$$

$$c_1(0) = 2a \tag{1.3.17}$$

has the two solutions

$$c_1(t) \equiv 2a, \tag{1.3.18}$$

$$c_1(t) = \begin{cases} 2a - \dfrac{k_1^2}{4} t^2, & 0 \leqq t \leqq \dfrac{2a^{\frac{1}{2}}}{k_1} \\ a e^{2 - k_1 a^{-\frac{1}{2}} t}, & \dfrac{2a^{\frac{1}{2}}}{k_1} \leqq t. \end{cases} \tag{1.3.19}$$

Lacking stronger results, we resort to the standard uniqueness theorem.

Theorem 1.3.2. If the vector valued function $\mathbf{f}(\xi)$ is Lipschitzian in $\tilde{\Gamma}(u_0)$ the initial value problem (1.3.1), (1.3.2) has at most one solution.

The Lipschitzian property can be formulated as

$$|\mathbf{f}(\xi) - \mathbf{f}(\bar{\xi})| \leqq b_1 |\xi - \bar{\xi}| \tag{1.3.20}$$

where the constant b_1 applies to the whole region $\tilde{\Gamma}(u_0)$. In particular, $\mathbf{f}$ satisfies Eq. (1.3.20) if it possesses continuous derivatives. The requirement of continuous derivatives has physical basis but is quite restrictive.

1.4. Equilibrium Points of the Kinetic Equations

Now that the existence of solutions for Eqs. (1.3.1), (1.3.2) has been established for all times, it is natural to inquire about their asymptotic behavior as $t \to \infty$. In the next section, it will be shown that if the rate functions, $f_1, \dots, f_R$ are compatible with the thermodynamic requirement of positive entropy production all the trajectories converge to a single equilibrium point. For rate functions satisfying Postulate 1.2.1 only, the phase space can be very complex and the trajectories need not all converge to a single equilibrium point and, in fact, need not converge at all. However, critical or equilibrium points do exist as we shall prove shortly.

Before proceeding to the main subject we shall make an innocuous refinement of the requirements of Postulate 1.2.1 characterizing the admissible rate functions. It has been assumed for mathematical convenience that the functions f_j are defined over the whole space E_{N+1} although only the positive orthant E_{N+1}^+ is of physical interest. In time dependent problems all state trajectories originating in $\Gamma(u_0)$ never leave this region so that the particular definition of f_j in $E_{N+1} - \Gamma(u_0)$ has no consequence. In the case of time independent problems, however, it is possible that some equilibrium points may lie in the region of negative state variables, $E_{N+1} - E_{N+1}^+$. Such *spurious* points result from the particular definition of the functions f_j in $E_{N+1} - E_{N+1}^+$ and do not have any physical significance.

Now the spurious equilibrium points can be ignored for all purposes except for the calculation of the rotation of the vector field $\tilde{\mathbf{f}}$ on the boundary of the manifold $\tilde{\Gamma}(u_0)$. The rotation, which gives useful information about the number and stability of equilibrium points, does not exist if an equilibrium point lies on the boundary of the manifold. This difficulty is circumvented by calculating the rotation on a sufficiently large sphere surrounding the manifold. The sphere, however, inevitably includes some regions of negative state variables and it must be insured that it does not include any spurious equilibrium point. This is achieved by the following special definition of the functions f_j.

Let $\mathbf{u}_0 \in E_{N+1}^+$ and let $\Gamma'(u_0)$ be the manifold defined by Eqs. (1.2.13), (1.2.14). Clearly, $\Gamma'(u_0)$ is the extension of $\Gamma(u_0)$ in E_{N+1}. For any point $\mathbf{u} \in \Gamma'(u_0)$ with some negative components denoted by c^-, we define

$$\sum_{j=1}^{R} v_{ij} f_j(c_1, \ldots, c^-, \ldots, T) = \sum_{j=1}^{R} v_{ij} f_j(c_1, \ldots, 0, \ldots, T) + (c_{i0} - c_i) d(\mathbf{u}) \quad (1.4.1)$$

$$i = 1, \ldots, N$$

where $f_j(c_1, \ldots, 0, \ldots, T)$ is obtained from $f_j(c_1, \ldots, c^-, \ldots, T)$ by replacing all negative components by 0 and $d(\mathbf{u})$ is the distance between the point $\mathbf{u}$ and the closed manifold $\Gamma(u_0) \subset E_{N+1}^+$. The distance function d is introduced to preserve the continuity of the functions f_j. Eqs. (1.4.1) have a unique solution in the unknowns $f_j(c_1, \ldots, c^-, \ldots, T)$ by virtue of Eqs. (1.2.9), (1.2.13) and the fact that the matrix $\boldsymbol{v}$ has rank R. It must be emphasized that the definition (1.4.1) applies only to points in the manifold $\Gamma'(u_0)$. Similar definitions of the functions f_j in $E_{N+1} - E_{N+1}^+$ will be used for the steady state problems of Sections 1.8, 2.5.

A *kinetic equilibrium point*, or state, is defined as a solution of the equations

$$f_j(c_1, \ldots, c_N, T) = 0 \qquad j = 1, \ldots, R. \quad (1.4.2)$$

We shall often omit the adjective kinetic and simply say an equilibrium point. It will be verified now that Eqs. (1.4.1) do not allow the existence of an equilibrium point in $\Gamma'(u_0)-\Gamma(u_0)$. Indeed, if at some a point $c_i<0$, a combination of Eqs. (1.4.1), (1.4.2) gives the contradiction $c_{i0}-c_i \leqq 0$. If $T<0$ a combination of Eqs. (1.4.1), (1.4.2), (1.2.5) gives $c_k=c_{k0}$, $k=1,\ldots,N$ so that by Eq. (1.2.14), $U(c_{10},\ldots,c_{N0},T)=U(c_{10},\ldots,c_{N0},0)$ which contradicts Eq. (1.2.7). We are ready now to show the existence of equilibrium points and at the same time calculate the rotation of the vector field $\tilde{\mathbf{f}}$.

Theorem 1.4.1. If the rate functions $f_1,\ldots,f_R$ and the internal energy function U satisfy the conditions of Postulate 1.2.1 then each invariant manifold $\Gamma(u_0)$ contains at least one equilibrium point.

Proof. If we define extents $\xi_1,\ldots,\xi_R$ by

$$c_i-c_{i0}=\sum_{j=1}^{R}\nu_{ij}\,\xi_j \tag{1.4.3}$$

then an equilibrium point is defined by the equation

$$\tilde{\mathbf{f}}(\boldsymbol{\xi})=0 \tag{1.4.4}$$

and must lie in $\tilde{\Gamma}(u_0)$, otherwise the corresponding point $\mathbf{u}$ would have some negative components, which is not possible.

We shall show that the vector fields $\tilde{\mathbf{f}}$ and $-\mathbf{I}$ are homotopic on a sphere $S=\{\boldsymbol{\xi}\colon |\boldsymbol{\xi}|=b\}$ surrounding the bounded region $\tilde{\Gamma}(u_0)$. For this purpose we consider the vector field

$$\mathscr{H}(s)=-s\,\mathbf{I}+(1-s)\,\tilde{\mathbf{f}} \tag{1.4.5}$$

which obviously satisfies $\mathscr{H}(0)=\tilde{\mathbf{f}}$, $\mathscr{H}(1)=-\mathbf{I}$. It remains to show that $\mathscr{H}(s)$ has no null points on S for any $0\leqq s\leqq 1$. If $\mathscr{H}(s)$ vanishes at a point $\boldsymbol{\xi}\in S$ then

$$\tilde{f}_j=\frac{1-s}{s}\,\xi_j \qquad j=1,\ldots,R \tag{1.4.6}$$

from which one obtains

$$\sum_{j=1}^{R}\nu_{ij}\tilde{f}_j=\frac{1-s}{s}\,(c_i-c_{i0}). \tag{1.4.7}$$

But $\boldsymbol{\xi}\in S$ implies that the corresponding state $\mathbf{u}$ has some negative components. If the negative component is $c_i<0$, by using Eq. (1.4.1) we can rewrite Eq. (1.4.7) as

$$0\leqq\sum_{j=1}^{R}\nu_{ij}f_j(c_1,\ldots,0,\ldots,T)=-(c_{i0}-c_i)\left[d(\mathbf{u})+\frac{1-s}{s}\right] \tag{1.4.8}$$

which is a contradiction. Similarly, if $T<0$, we arrive at the contradiction $U(c_{10},\ldots,c_{N0},0)=U(c_{10},\ldots,c_{N0},T)$. The vector fields $\tilde{\mathbf{f}}$ and $-\mathbf{I}$ as homotopic have the same rotation $(-1)^R$, therefore Eqs. (1.4.2) have one at least solution in $\Gamma(u_0)$. Some results on the number and stability of the equilibrium points will be derived in Section 1.7. Fig. 1.4.1 shows schematically the homotopic vector fields $\tilde{\mathbf{f}}$, $-\mathbf{I}$ for the case of two chemical reactions.

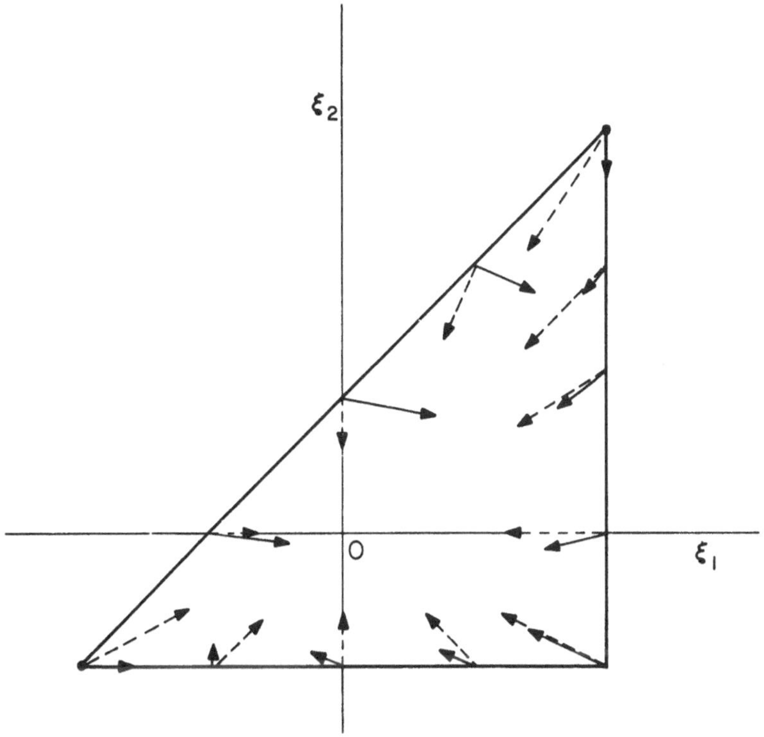

Fig. 1.4.1. Schematic representation of the vector fields $-\mathbf{I}, \tilde{\mathbf{f}}$ on $\partial\tilde{\Gamma}(u_0)$

1.5. Convergence to the Equilibrium State

We shall now formulate certain conditions which insure the convergence of all trajectories, within a given manifold, to a single equilibrium point*. These conditions, if interpreted thermodynamically, express the consistency between thermodynamics and kinetics.

Postulate 1.5.1. There exists a function $S(c_1, \ldots, c_N, T)$ which along with U has continuous first partial derivatives in the interior of the region E^+_{N+1}. The quantities

$$A_j(c_1, \ldots, c_N, T) = -\sum_{i=1}^{N} \nu_{ij}\, \mu_i(c_1, \ldots, c_N, T) \qquad j = 1, \ldots, R \tag{1.5.1}$$

* Attention is confined here to classical equilibrium in BOWEN'S sense [11].

where

$$\underset{i}{\mu}(c_1, \ldots, c_N, T) = T\left(\frac{\partial U}{\partial T}\right)^{-1}\left[\frac{\partial U}{\partial T}\frac{\partial S}{\partial c_i} - \frac{\partial U}{\partial c_i}\frac{\partial S}{\partial T}\right] \qquad i=1, \ldots, n \qquad (1.5.2)$$

satisfy the following requirements:

(i) The manifold Ω defined by the equations

$$A_j(c_1, \ldots, c_N, T) = 0 \qquad j=1, \ldots, R \qquad (1.5.3)$$

is identical to the manifold Ω' defined by the equations

$$f_j(c_1, \ldots, c_N, T) = 0 \qquad j=1, \ldots, R. \qquad (1.5.4)$$

(ii)

$$\sum_{j=1}^{R} f_j A_j \geqq 0 \qquad (1.5.5)$$

the equality applying only for $\mathbf{u} \in \Omega$.

(iii) Each invariant manifold $\Gamma(u_0)$ intersects Ω at a single point $\mathbf{u}^* = (c_1^*, \ldots, c_N^*, T^*)$, the equilibrium point of $\Gamma(u_0)$.

(iv) Within a given invariant manifold

$$S(c_1^*, \ldots, c_N^*, T^*) - S(c_1, \ldots, c_N, T) \geqq 0 \qquad (1.5.6)$$

with equality holding only at $\mathbf{u} = \mathbf{u}^*$.

An invariant manifold $\Gamma(u^*)$ containing the equilibrium point $\mathbf{u}^*$, can be described by the following extent variables

$$c_i - c_i^* = \sum_{j=1}^{R} \nu_{ij}\,\xi_j \qquad i=1, \ldots, N. \qquad (1.5.7)$$

Within $\Gamma(u^*)$, the functions S, A_j depend on the extents alone so that we may define

$$\tilde{S}(\xi_1, \ldots, \xi_R) = S(c_1, \ldots, c_N, T), \qquad (1.5.8)$$

$$\tilde{A}_j(\xi_1, \ldots, \xi_R) = A_j(c_1, \ldots, c_N, T) \qquad (1.5.9)$$

and write Eq. (1.5.6) as

$$\tilde{S}(0, \ldots, 0) - \tilde{S}(\xi_1, \ldots, \xi_R) \geqq 0. \qquad (1.5.10)$$

It may be easily verified that

$$\frac{\partial \tilde{S}}{\partial \xi_j} = \frac{\tilde{A}_j}{T} \qquad (1.5.11)$$

therefore

$$\frac{d}{dt}\left[\tilde{S}(0, \ldots, 0) - \tilde{S}(\xi_1, \ldots, \xi_R)\right] = -\frac{1}{T}\sum_{j=1}^{R} \tilde{A}_j \tilde{f}_j \leqq 0. \qquad (1.5.12)$$

The following theorem follows directly from Postulate 1.5.1 and Eq. (1.5.12).

Theorem 1.5.1. In any given invariant manifold $\tilde{\Gamma}(u^*)$, the system of Eqs. (1.3.1), (1.3.2) is asymptotically stable in the large, i.e., for all trajectories emanating from the interior of $\tilde{\Gamma}(u^*)$

$$\lim_{t\to\infty} \xi(t)=0. \tag{1.5.13}$$

In the thermodynamic interpretation of Postulate 1.5.1, S is the entropy, μ_i the chemical potentials, and A_j the affinities. The manifolds Ω, Ω' can be called the thermodynamic and kinetic manifolds. The quantity $(\sum_j A_j f_j)/T$ is the entropy production so that conditions (i), (ii) of Postulate 1.5.1 are the conditions of consistency between kinetics and the second law of thermodynamics. Conditions (iii), (iv) will be proved in the next section for ideal gas mixtures. In general, however, they should be taken as additional assumptions. To our knowledge, no experimental system has been reported which does not satisfy these conditions.

Postulate 1.5.1 states that $S(\mathbf{u}^*)-S(\mathbf{u})$ is a Liapounov function for an isolated system. Other types of closed systems have, in accordance to the second law of thermodynamics, their own potential or Liapounov functions. For example, an isothermal-isobaric system has the Gibbs free energy $G(\mathbf{u})-G(\mathbf{u}^*)$, an isothermal-isochoric system has the Helmholtz free energy $A(\mathbf{u})-A(\mathbf{u}^*)$.

Open systems do not in general possess thermodynamic potential functions, a fact which often manifests itself with the existence of more than one equilibrium points in the same invariant manifold. Open systems still satisfy the condition of positive entropy production, Eq. (1.5.5). Some interesting implications of the positive entropy production for the steady states of open distributed systems are discussed in Section 2.8.

Empirical rate functions will not generally be consistent with thermodynamics. In this case the asymptotic behavior of the system can be studied by means of Liapounov functions having the properties of Postulate 1.5.1 but not having thermodynamic significance. These non-thermodynamic Liapounov functions would also be useful when thermodynamic expressions for the entropy are not available.

There is considerable literature relevant to the subject of the present section. The textbooks of PRIGOGINE and DEFAY [35] and of DENBIGH [14] present well the thermodynamics of chemical equilibrium. The textbook of DE GROOT and MAZUR [13] on irreversible thermodynamics treats in detail the subject of entropy production. The subject of consistency of kinetics with thermodynamics has been treated among others by ARIS [5], BOWEN [11], DENBIGH [14]. Finally, the subject of Liapounov functions and approach to equilibrium has been treated by WEI [40].

1.6. Ideal Gas Mixtures

Let us consider an ideal gas mixture of N species, subject to R independent reaction with rates $f_1, \ldots, f_R$. The internal energy function is taken as

$$U = \sum_{i=1}^{N} c_i \int_0^T c_{vi}(T')\, dT' \equiv U_0 \tag{1.6.1}$$

where $c_{vi}(T) > 0$ for all T.

The time evolution of the system is described by Eq. (1.6.1) and by the equation

$$\frac{dc_i}{dt} = \sum_{j=1}^{R} \nu_{ij} f_j. \tag{1.6.2}$$

In ideal gas mixtures the entropy is given by

$$S(c_1, \ldots, c_N, T) = \sum_{i=1}^{N} c_i S_i - R_g \sum_{i=1}^{N} c_i \ln \frac{c_i}{c_T} \tag{1.6.3}$$

where

$$S_i = S_{i0} - R_g \ln \frac{p}{p_0} + \int_{T_0}^{T} [c_{vi}(T') + R_g] \frac{dT'}{T'}. \tag{1.6.4}$$

R_g is the gas constant, p the pressure given by

$$p = R_g c_T T, \tag{1.6.5}$$

$$c_T = \sum_{i=1}^{N} c_i \tag{1.6.6}$$

and p_0, T_0, S_{i0} are constants. The chemical potentials and affinities are calculated from Eq. (1.6.3) as

$$\mu_i = \mu_i^0 + R_g T \ln(c_i R_g T), \tag{1.6.7}$$

$$A_j = -\Delta F_j^0 - R_g T \sum_{i=1}^{N} \nu_{ij} \ln(c_i R_g T) \tag{1.6.8}$$

where μ_i^0, ΔF_j^0 are functions of the temperature only, given by

$$\begin{aligned} \mu_i^0 = &- T(S_{i0} - R_g + R_g \ln p_0) - \\ &- T \int_{T_0}^{T} [c_{vi}(T') + R_g]\, dT' + \int_0^T c_{vi}(T')\, dT', \end{aligned} \tag{1.6.9}$$

$$\Delta F_j^0 = \sum_{i=1}^{N} \nu_{ij} \mu_i^0. \tag{1.6.10}$$

The equilibrium manifold Ω is defined by $A_j = 0$ which gives

$$\sum_{i=1}^{N} \nu_{ij} \ln(c_i R_g T) = -\frac{\Delta F_j^0}{R_g T} \qquad j = 1, \ldots, R, \tag{1.6.11}$$

or

$$\prod_{i=1}^{N} c_i^{\nu_{ij}} = K_j(T) \qquad j=1, \ldots, R \tag{1.6.12}$$

where

$$K_j(T) = (R_g T)^{-\Delta \nu_j} e^{-\frac{\Delta F_j^0}{R_g T}}, \tag{1.6.13}$$

$$\Delta \nu_j = \sum_{i=1}^{N} \nu_{ij}. \tag{1.6.14}$$

The following two theorems show that the function S defined by Eq.(1.6.3) satisfies the requirements (iii) and (iv) of Postulate 1.5.1.

Theorem 1.6.1. The maximum of $\tilde{S}(\xi_1, \ldots, \xi_R)$ in any given invariant manifold $\tilde{\Gamma}(u_0)$ is attained in the interior of the manifold.

Proof. We recall that $\tilde{\Gamma}(u_0)$, the image of $\Gamma(u_0)$ in the $\boldsymbol{\xi}$ space, is a closed bounded and connected set. The boundary $\partial\tilde{\Gamma}(u_0)$ consists of all points where at least one of the variables $c_1, \ldots, c_N, T$ vanishes. First, we verify that the maximum of $\tilde{S}$ cannot be attained at a point $\boldsymbol{\xi}_T$ where $T=0$, because as $\boldsymbol{\xi} \to \boldsymbol{\xi}_T$, $\tilde{S} \to -\infty$. Next we consider a point $\boldsymbol{\xi}_\alpha$ where $c_\alpha = 0$ and $c_i > 0$, $i \neq \alpha$. The point $\boldsymbol{\xi}_\alpha$ lies on the hyperplane

$$\sum_{j=1}^{R} \nu_{\alpha j} \xi_j + c_{\alpha 0} = 0. \tag{1.6.15}$$

The outward unit normal on $\partial\tilde{\Gamma}(u_0)$ at the point $\boldsymbol{\xi}_\alpha$ is $-|\boldsymbol{\nu}_\alpha|^{-1} \boldsymbol{\nu}_\alpha$ where $\boldsymbol{\nu}_\alpha = (\nu_{\alpha 1}, \ldots, \nu_{\alpha R})$. The directional derivative of $\tilde{S}$ along that normal is

$$\frac{\partial S}{\partial n} = -\frac{|\boldsymbol{\nu}_\alpha|^{-1}}{T} \sum_{j=1}^{R} \nu_{\alpha j} \tilde{A}_j. \tag{1.6.16}$$

In the interior of $\tilde{\Gamma}(u_0)$ and near the point $\boldsymbol{\xi}_\alpha$ the quantity $\tilde{A}_j$ has the form

$$\tilde{A}_j = -R_g T \nu_{\alpha j} \ln c_\alpha + r_j \tag{1.6.17}$$

where r_j is negligible compared to the first term. Introducing this expression in Eq.(1.6.16) there is obtained

$$\frac{\partial \tilde{S}}{\partial \alpha} = R_g |\boldsymbol{\nu}_\alpha| \ln c_\alpha + r_\alpha < 0 \tag{1.6.18}$$

where r_α is a negligible term. This shows that $\tilde{S}$ cannot attain its maximum at a point like $\boldsymbol{\xi}_\alpha$.

Next, consider a point $\boldsymbol{\xi}_{\alpha\beta}$ at which $c_\alpha = 0$, $c_\beta = 0$, $c_i \neq 0$, $i \neq \alpha, \beta$. This point lies on the intersection of the hyperplanes Π_α and Π_β defined by

$$\sum_{j=1}^{R} \nu_{\alpha j} \xi_j + c_{\alpha 0} = 0, \tag{1.6.19}$$

$$\sum_{j=1}^{R} \nu_{\beta j} \xi_j + c_{\beta 0} = 0. \tag{1.6.20}$$

The function $\tilde{S}$ limited in the hyperplane Π_α is given by

$$\tilde{S}=\sum_{i\neq\alpha}^{N} c_i S_i - R_g \sum_{i\neq\alpha}^{N} c_i \ln \frac{c_i}{c_T}. \tag{1.6.21}$$

Consider the unit vector γ parallel to Π_α and perpendicular to the intersection of Π_α and Π_β pointing in the direction of decreasing c_β i.e.

$$\nu_\beta \gamma < 0. \tag{1.6.22}$$

Within the hyperplane Π_α and in the neighborhood of $\xi_{\alpha\beta}$ the directional derivative of $\tilde{S}$ along γ is easily calculated as

$$\frac{\partial S}{\partial n} = -R_g \nu_\beta \gamma \ln c_\beta + r_{\alpha\beta} < 0 \tag{1.6.23}$$

where $r_{\alpha\beta}$ is negligible. It follows again that $\tilde{S}$ cannot attain its maximum at $\xi_{\alpha\beta}$. A similar argument can be given for points that lie on the intersection of more than two hyperplanes. It follows, therefore, that the maximum of $\tilde{S}$ is attained in the interior of $\tilde{\Gamma}(u_0)$.

The following theorem has been proved by ARIS [5].

Theorem 1.6.2. The equations:

$$\tilde{A}_j = 0 \qquad j=1, \ldots, R \tag{1.6.24}$$

have a unique solution in the interior of $\tilde{\Gamma}(u_0)$.

Proof. The existence of a solution follows from the previous theorem since at the point at which $\tilde{S}$ attains its maximum all the derivatives $\tilde{A}_j/T$ must vanish. Uniqueness will be proved by showing that the matrix with jk^{th} element:

$$J_{jk} = -\frac{\partial}{\partial \xi_k}\left(\frac{\tilde{A}_j}{R_g T}\right) \tag{1.6.25}$$

is positive definite in the interior of $\tilde{\Gamma}(u_0)$. From Eq. (1.6.8) there is obtained

$$J_{jk} = a_{jk} + b_{jk} \tag{1.6.26}$$

where:

$$a_{jk} = \frac{\partial}{\partial \xi_k} \sum_{i=1}^{N} \nu_{ij} \ln c_i = \sum_{i=1}^{N} \frac{\nu_{ij}\nu_{ik}}{c_i}, \tag{1.6.27}$$

$$b_{jk} = \frac{\partial}{\partial \xi_k}\left(\frac{\Delta F_j^0}{R_g T}\right) = \frac{\sum_{i=1}^{N} d_{ij} d_{ik}}{R_g T^2 \sum_{i=1}^{N} c_i c_{iv}(T)}, \tag{1.6.28}$$

$$d_{ij} = \sum_{i=1}^{N} \nu_{ij} \int_0^T c_{vi}(T')\, dT'. \tag{1.6.29}$$

The matrix (a_{jk}) is positive definite because for any vector $(\rho_1, \ldots, \rho_R) \neq 0$

$$\sum_{j=1}^{R} \sum_{k=1}^{R} a_{jk} \rho_j \rho_k = \sum_{i=1}^{N} \frac{1}{c_i} \left(\sum_{j=1}^{R} \nu_{ij} \rho_j \right)^2 > 0 \tag{1.6.30}$$

due to the fact that the R columns of the matrix (ν_{ij}) are linearly independent. The matrix b_{jk} is positive semidefinite because:

$$\sum_{j=1}^{R} \sum_{k=1}^{R} b_{jk} \rho_j \rho_k = \frac{\sum_{i=1}^{N} \left(\sum_{j=1}^{R} d_{ij} \rho_j \right)^2}{R_g T^2 \sum_{i=1}^{N} c_i c_{vi}(T)} \geqq 0. \tag{1.6.31}$$

Finally, the matrix J_{jk} is also positive definite, as the sum of a positive definite and a positive semidefinite matrix.

We shall close this section with a few examples of rate functions which are consistent with thermodynamics.

(i)
$$f_j = A_j. \tag{1.6.32}$$

These functions are not defined on the boundary of E_{N+1}^+.

(ii)
$$f_j = \frac{e^{-|A_j|} + |A_j| - 1}{A_j} \tag{1.6.33}$$

these functions are defined on the entire region E_{N+1}^+.

These two examples do not correspond to any experimental system. In the case of ideal gases, some reactions have been experimentally observed to have rates of the form

$$f_j = k_j^+(T) \prod_{i=1}^{N} c_i^{\alpha_{ij}^-} - k_j^-(T) \prod_{i=1}^{N} c_i^{\alpha_{ij}^+} \tag{1.6.34}$$

where $k_j^+(T)$, $k_j^-(T)$ are positive functions such that

$$\frac{k_j^+(T)}{k_j^-(T)} = K_j(T) = e^{-\frac{\Delta F_j^0}{R_g T}} (R_g T)^{-\Delta \nu_j} \tag{1.6.35}$$

and

$$\alpha_{ij}^+ = \tfrac{1}{2}(|\nu_{ij}| + \nu_{ij}), \qquad \alpha_{ij}^- = \tfrac{1}{2}(|\nu_{ij}| - \nu_{ij}). \tag{1.6.36}$$

For example, the rate of the reaction

$$O_2 + 2NO - 2NO_2 = 0 \tag{1.6.37}$$

has been observed as

$$f = k^+(T)\, c_{NO}^2\, c_{O_2} - k^-(T)\, c_{NO_2}^2. \tag{1.6.38}$$

The rate functions given by Eq. (1.6.34) clearly satisfy condition (i) of Postulate (1.5.1). It is interesting to rewrite f_j as

$$f_j = k_j^-(T)\left(e^{\frac{A_j}{R_g T}} - 1\right)\prod_{i=1}^{N} c_i^{\alpha_i^+ j}. \qquad (1.6.39)$$

This form shows a special relationship between rate and affinity. VAN RYSSELBERGHE [39] discusses in some detail the relationship between rates and affinities.

From Eq. (1.6.39) it is obvious that the quantity $A_j f_j$ is positive at all points outside the equilibrium manifold. This means that in chemical reaction systems with kinetics given by Eq. (1.6.34), each reaction individually leads to positive entropy production, a result beyond the requirements of Postulate 1.5.1.

1.7. The Number and Stability of Equilibrium States in Closed Systems

In Section 1.4 the topological concept of rotation was used to prove the existence of equilibrium states. When the reaction kinetics are not restricted by Postulate 1.5.1, each invariant manifold may include more than one equilibrium states and it is interesting to obtain information about the number and stability of these states. In the present section we shall use one more topological concept, the index of a fixed point, to show that the equilibrium states are odd in number, $2m+1$, among which m at least are unstable. As in the preceding sections, the discussion concerns isolated systems, but extension to other closed systems should not present difficulties.

An equilibrium state or point $\xi \in \tilde{\Gamma}(u_0)$ is defined as a solution of the equation

$$\tilde{\mathbf{f}}(\xi) = 0 \qquad (1.7.1)$$

which will be written as

$$\xi - (\xi - \tilde{\mathbf{f}}(\xi)) = 0 \qquad (1.7.2)$$

to conform with the notation of the Appendix. We shall assume that the functions $\tilde{f}_1, \ldots, \tilde{f}_R$ are extended in E_R by Eq. (1.4.1) and that the functions $f_1, \ldots, f_R$ are continuously differentiable in E_R^+. This implies that the functions $\tilde{f}_1, \ldots, \tilde{f}_R$ are continuously differentiable in $E_R - \partial\tilde{\Gamma}(u_0)$ and have one sided derivatives on the closed surface $\partial\tilde{\Gamma}(u_0)$.

According to Theorems A.5, A.6, the calculation of the index of a point $\boldsymbol{\xi}^*$ satisfying Eq. (1.7.2) is based on the eigenvalue problem

$$\boldsymbol{\phi} = \lambda(\boldsymbol{\phi} - \mathbf{A}(\boldsymbol{\xi}^*)\,\boldsymbol{\phi}) = 0 \qquad (1.7.3)$$

or equivalently

$$\mathbf{A}(\boldsymbol{\xi}^*)\,\boldsymbol{\phi} = \lambda'\boldsymbol{\phi} \qquad (1.7.4)$$

where $\boldsymbol{\phi}=(\phi_1, \ldots, \phi_R)$ and $\mathbf{A}(\xi)$ is the $R\times R$ matrix

$$A_{jk}(\xi)=\frac{\partial \tilde{f}_j}{\partial \xi_k}. \tag{1.7.5}$$

The index of the point ξ^* is given by $(-1)^\beta$ where β is the sum of multiplicities of the eigenvalues $\lambda\in(0, 1)$ of Eq. (1.7.3) or, equivalently, of the eigenvalues $\lambda'\in(-\infty, 0)$ of the matrix $\mathbf{A}(\xi^*)$.

We are ready to prove

Theorem 1.7.1. If the matrix $\mathbf{A}(\xi^*)$ is nonsingular for all equilibrium points $\xi^*\in\tilde{\Gamma}(u_0)$ then the number of these points is odd, $n=2m+1$, among which $m+1$ have index $(-1)^R$ and the remaining m have index $(-1)^{R+1}$.

Proof. By applying Theorems A.5, A.6 it is found that the index of any fixed point of the operator $\mathbf{I}-\tilde{\mathbf{f}}$ is ± 1. Furthermore, the sum of the indices of the fixed points is equal to the rotation of the operator $\mathbf{I}-\mathbf{f}$ on any sphere surrounding $\tilde{\Gamma}(u_0)$. This rotation has been calculated as $(-1)^R$ in the proof of Theorem 1.4.1. Thus the number of fixed points is odd, $2m+1$, among which $m+1$ have index $(-1)^R$ and the remaining m have index $(-1)^{R+1}$.

Let us now examine the exceptional case of an initial point $\bar{\mathbf{u}}_0$ and an equilibrium point $\xi^*\in\tilde{\Gamma}(\bar{u}_0)$ such that the matrix $\mathbf{A}(\xi^*)$ is singular. Recalling the one to one correspondence between $\mathbf{u}$ and ξ, Eqs. (1.2.21) to (1.2.23), we see that the equations

$$f_j(c_1(\xi; \mathbf{u}_0), \ldots, c_N(\xi; \mathbf{u}_0), T(\xi; \mathbf{u}_0))=0 \qquad j=1, \ldots, R \tag{1.7.6}$$

define implicitly ξ as a function of $\mathbf{u}_0$ and have the solution $\xi=\xi^*$ at $\mathbf{u}_0=\bar{\mathbf{u}}_0$. The Jacobian matrix of the transformation defined by Eq. (1.7.6) is $\mathbf{A}(\xi^*)$. When $\mathbf{A}(\xi^*)$ is nonsingular, Eq. (1.7.6) define a unique and continuous solution $\xi=\xi(\mathbf{u}_0)$ at the neighborhood of $\bar{\mathbf{u}}_0$. When $\mathbf{A}(\xi^*)$ is singular, Eqs. (1.7.6) do not define uniquely ξ as a function of $\mathbf{u}_0$. This idea can be elaborated somewhat further. Let $L\subset E^+_{N+1}$ be the N-dimensional surface of points $\mathbf{u}_0$ such that $\mathbf{A}(\xi(\mathbf{u}_0))$ is singular and let V_0 be a small ball centered at $\bar{\mathbf{u}}_0$. If $\mathbf{u}_0\in V_0-L$, Eqs. (1.7.6) do not have a unique solution $\xi(\mathbf{u}_0)$ in the neighborhood of ξ^*. Usually, when $\mathbf{u}_0$ lies from the one side of the surface L, Eq. (1.7.6) has no solution and when $\mathbf{u}_0$ lies from the other side, Eq. (1.7.6) has two solutions, one with index $+1$ the other with index -1. In such cases the index of the point $\xi^*=\xi(\bar{\mathbf{u}}_0)$ is 0.

Stability

An equilibrium state $\mathbf{u}^*\in\Gamma(u^*)$ will be called stable if given any $\varepsilon>0$ there exists a $\delta>0$ such that $|\mathbf{u}_0'-\mathbf{u}^*|<\delta$ implies that the solution $\mathbf{u}(t)$ of the initial value problem

$$\frac{dc_i}{dt}=\sum_{j=1}^{R}\nu_{ij}f_j \qquad i=1,\dots,N, \tag{1.7.7}$$

$$U(c_1,\dots,c_N,T)=U(c'_{10},\dots,c'_{N0},T'_0), \tag{1.7.8}$$

$$t=0:\ c_i=c'_{i0}, \qquad T=T' \tag{1.7.9}$$

satisfies

$$|\mathbf{u}(t)-\mathbf{u}^*|<\varepsilon, \qquad 0\leqq t. \tag{1.7.10}$$

We wish first to show that for the purpose of stability analysis there is no loss of generality in considering perturbations lying in the manifold of the equilibrium state, i.e., $\mathbf{u}'_0\in\Gamma(u^*)$. To this end we define extents $\xi_1,\dots,\xi_R$ by

$$c_i-c_i^*=\sum_{j=1}^{R}\nu_{ij}\,\xi_j \qquad i=1,\dots,N \tag{1.7.11}$$

and thus define $\mathbf{u}$ as a function of $\mathbf{u}^*$ and $\boldsymbol{\xi}$, i.e., $\mathbf{u}=\mathbf{u}(\boldsymbol{\xi};\mathbf{u}^*)$. We also set

$$A_{jk}(\boldsymbol{\xi};\mathbf{u}^*)=\frac{\partial f_j(\mathbf{u}(\boldsymbol{\xi};\mathbf{u}^*))}{\partial\xi_k} \tag{1.7.12}$$

to emphasize the dependence on both $\boldsymbol{\xi}$ and $\mathbf{u}^*$. Now according to the discussion following Eq. (1.7.6), if the Jacobian matrix $A_{jk}(0,\mathbf{u}^*)$ is non-singular, there exists a unique equilibrium state $\mathbf{u}^{*\prime}\in\Gamma(u'_0)$ satisfying $|\mathbf{u}^{*\prime}-\mathbf{u}^*|<\varepsilon/2$ provided $|\mathbf{u}^*-\mathbf{u}'_0|$ is sufficiently small. Thus Eq. (1.7.10) will be satisfied if

$$|\mathbf{u}(t)-\mathbf{u}^{*\prime}|<\frac{\varepsilon}{2}. \tag{1.7.13}$$

This shows that the stability of $\mathbf{u}^*\in\Gamma(u^*)$ under a perturbation $\mathbf{u}'_0$ is equivalent to the stability of $\mathbf{u}^{*\prime}\in\Gamma(u'_0)$ under the perturbation $\mathbf{u}'_0$. The latter problem, however, is described by the reduced equations

$$\frac{d\boldsymbol{\zeta}}{dt}=\mathbf{f}(\mathbf{u}(\boldsymbol{\zeta};\mathbf{u}'_0)), \tag{1.7.14}$$

$$\boldsymbol{\zeta}(0)=0 \tag{1.7.15}$$

where the extents ζ_j are defined by

$$c_i-c_i^{*\prime}=\sum_{j=1}^{R}\nu_{ij}\,\zeta_j. \tag{1.7.16}$$

Now the local stability of Eq. (1.7.14) is determined by the eigenvalues of the matrix $\mathbf{A}(0,\mathbf{u}^{*\prime})$ or, equivalently, by the eigenvalues of the matrix

$\mathbf{A}(0, \mathbf{u}^*)$, for $|\mathbf{u}^{*\prime} - \mathbf{u}^*|$ can be chosen arbitrarily small. We can de-emphasize now the dependence of the functions $\mathbf{f}, \mathbf{A}$ on $\mathbf{u}^*$ and return to a simpler and more familiar notation. Let $\mathbf{u}^* \in \Gamma(u_0)$ and

$$c_i - c_{i0} = \sum_{j=1}^{R} v_{ij} \xi_j, \tag{1.7.17}$$

$$c_i^* - c_{i0} = \sum_{j=1}^{R} v_{ij} \xi_j^*, \tag{1.7.18}$$

$$\tilde{\mathbf{f}}(\xi) = \mathbf{f}(\mathbf{u}(\xi; \mathbf{u}_0)), \tag{1.7.19}$$

$$A_{jk}(\xi) = \frac{\partial \tilde{f}_j}{\partial \xi_k}. \tag{1.7.20}$$

The stability of the equilibrium state $\boldsymbol{\xi}^*$ is completely determined by the eigenvalues of the matrix $\mathbf{A}(\boldsymbol{\xi}^*)$. By LIAPOUNOV's theorem, a sufficient condition for the local asymptotic stability of the equilibrium state $\boldsymbol{\xi}^*$ is that all eigenvalues of the matrix $\mathbf{A}(\boldsymbol{\xi}^*)$ have negative real parts. Conversely, if the matrix $\mathbf{A}(\boldsymbol{\xi}^*)$ has one or more eigenvalues with positive real parts, the equilibrium point $\boldsymbol{\xi}^*$ is unstable. Finally, when the matrix $\mathbf{A}(\boldsymbol{\xi}^*)$ has points with zero real parts no definitive statement about stability can be made.

To obtain the relationship between the index and the stability of an equilibrium point let us consider first an odd number R of chemical reactions. Then the matrix $\mathbf{A}(\boldsymbol{\xi}^*)$ has an odd number of real eigenvalues (counting each with its multiplicity) due to the fact that complex eigenvalues appear in pairs. An index $+1$ implies, according to Theorem 1.7.1, an even number of eigenvalues in $(-\infty, 0)$ and hence an odd number of eigenvalues in $(0, \infty)$ so that the equilibrium state $\boldsymbol{\xi}^*$ is unstable. An index -1 implies an odd number of eigenvalues in $(-\infty, 0)$ and hence an even number of eigenvalues in $(0, \infty)$. If $R=1$, $\boldsymbol{\xi}^*$ is stable but if $R=3, 5, \ldots$ no conclusion about stability can be made, for the number of eigenvalues in $(0, \infty)$ can be either zero or some other even number. In any case, if the number of equilibrium points is $2m+1$, at least m of them are unstable.

When R is even, the matrix $\mathbf{A}(\boldsymbol{\xi}^*)$ has an even number of real eigenvalues. An index -1 implies an odd number of eigenvalues in $(-\infty, 0)$ and an odd number in $(0, \infty)$, hence the equilibrium state $\boldsymbol{\xi}^*$ is unstable. An index $+1$ does not allow a conclusion about stability. Again, at least m among the $2m+1$ equilibrium states are unstable. We have proved the following theorem.

Theorem 1.7.2. An equilibrium state $\boldsymbol{\xi}^*$ such that the matrix $\mathbf{A}(\boldsymbol{\xi}^*)$ is nonsingular, is unstable if its index γ satisfies $(-1)^R \gamma < 0$. If the matrix $\mathbf{A}(\boldsymbol{\xi}^*)$ is nonsingular for all the equilibrium points of a given manifold, the number of these points is odd, $2m+1$, among which m at least are unstable. In the case of one chemical reaction ($R=1$), m of the equilibrium points are unstable and the remaining $m+1$ are stable. Thus for $R=1, m=0$ the unique equilibrium state is always stable.

1.8. Uniform Open Systems

The previous sections were concerned with isolated chemical reaction systems but most of the results are extendable to other uniform closed systems. Open systems on the other hand present a distinct behavior as

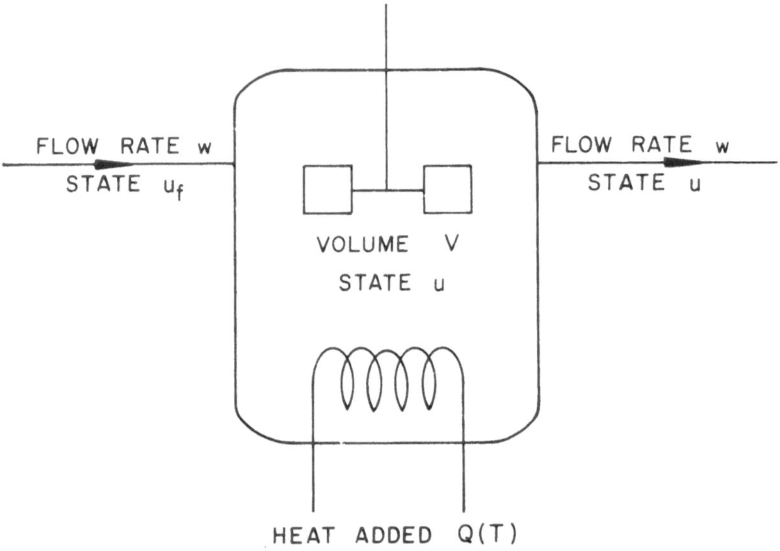

Fig. 1.8.1. A stirred-tank reactor

a result of their exchanging both energy and mass with their surroundings. Our treatment of open systems will be limited to a simple open system, known in chemical engineering as a continuous stirred tank reactor, or backmix reactor. The present section will contain the proof of existence of time dependent and time independent solutions for the appropriate differential equations. Some results on the number and stability of time independent solutions will be derived in the next section.

Fig. 1.8.1 shows schematically a continuous stirred tank reactor of constant volume V. w is the constant volumetric flow rate of input and output streams and $\mathbf{u}_f=(c_{1f}, \ldots, c_{Nf}, \mathrm{T}_f)$ is the input or feed state variables, which are given functions of time. As a result of the mixing, the state variables $\mathbf{u}=(c_1, \ldots, c_N, \mathrm{T})$ have the same values in the reactor and in the output stream. In addition to the energy carried by the input and output streams, the reactor can exchange heat with its surroundings, in an amount Q per unit time by means of a coil or a jacket or some other heat exchanging arrangement.

The conservation equations for the chemical species can be written immediately as

$$\frac{dc_i}{dt}=\frac{1}{\theta}(c_{if}-c_i)+\sum_{j=1}^{R}\nu_{ij}f_j \qquad i=1,\dots,N \tag{1.8.1}$$

where $\theta=V/w$ is called the *holding time* of the reactor. As the exact form of the energy equation is rather cumbersome, involving many thermodynamic functions, in our model we shall adopt the following commonly used simplified form

$$\frac{dT}{dt}=\frac{1}{\theta}(T_f-T)-\frac{1}{C_p}\sum_{j=1}^{R}\Delta H_j f_j+\frac{Q(T)}{VC_p} \tag{1.8.2}$$

where the quantity $C_p=\sum c_{pi}c_i$ and the heats of reactions

$$\Delta H_j=\sum_{i=1}^{N}\nu_{ij}H_i \tag{1.8.3}$$

are taken as constants. With no further loss of generality we can consider the partial molar enthalpies H_i as constants. The initial values of the state variables in the reactor will be denoted by $\mathbf{u}_0=(c_{10},\dots,c_{N0},T_0)$, i.e.

$$t=0:\ c_i=c_{i0} \qquad (i=1,\dots,N),\ T=T_0. \tag{1.8.4}$$

We wish to show that the initial value problem (1.8.1), (1.8.2), (1.8.4) has a solution for all times. The existence proof is easily carried out by the fixed point method of Section 1.3. The first task is to obtain a priori bounds for $c_1,\dots,c_N,T$. Multiplying Eq. (1.8.1) by γ_{il}, a quantity defined by Eq. (1.2.9), and summing over i there is obtained

$$\frac{da_l}{dt}+\frac{1}{\theta}a_l=\frac{1}{\theta}a_{lf} \tag{1.8.5}$$

where

$$a_l=\sum_{i=1}^{N}\gamma_{il}c_i, \tag{1.8.6}$$

$$a_{lf}=\sum_{i=1}^{N}\gamma_{il}c_{if}, \qquad a_{l_0}=\sum_{i=1}^{N}\gamma_{il}c_{i0}. \tag{1.8.7}$$

Eq. (1.8.5) gives by integration

$$a_l=a_{l_0}e^{-\frac{t}{\theta}}+\frac{1}{\theta}\int_0^t a_{lf}(\tau)e^{-\frac{t-\tau}{\theta}}d\tau. \tag{1.8.8}$$

Considering bounded inputs

$$|a_l(t)|\leqq a_{lM} \tag{1.8.9}$$

we obtain from Eq. (1.8.8)

$$a_l(t) < a_{l_0} + a_{lM}. \tag{1.8.10}$$

In particular, by setting $\gamma_{il} = \beta_{il} \geqq 0$, where β_{il} is the number of atoms $\mathscr{A}_l$ in the molecular species $\mathscr{M}_i$, we obtain

$$\sum_{i=1}^{N} \beta_{il}\, c_i(\mathrm{t}) < a_{l_0} + a_{lM}. \tag{1.8.11}$$

The last equation shows that the concentrations $c_1, \ldots, c_N$ are subject to a priori bounds. To obtain an a priori bound for the temperature T we shall assume first that the heat $Q(T)$ added to the reactor has the physically meaningful properties

$$0 < Q(0), \tag{1.8.12}$$

$$Q(T) \leqq Q_M, \qquad 0 \leqq T \tag{1.8.13}$$

where Q_M is a constant. Next we form the linear combination of Eqs. (1.8.1), (1.8.2)

$$\frac{dT_1}{dt} + \frac{1}{\theta} T_1 = \frac{1}{\theta} T_{1f} + \frac{Q(T)}{VC_p} \tag{1.8.14}$$

where

$$T_1 = T + \frac{1}{C_p} \sum_{i=1}^{N} H_i c_i, \tag{1.8.15}$$

$$T_{1f} = T_f + \frac{1}{C_p} \sum_{i=1}^{N} H_i c_{if}. \tag{1.8.16}$$

The a priori bound for T is obtained simply by integrating Eq. (1.8.14).

To apply the fixed point method we consider the Banach space $\mathscr{B}_c$ of vector valued functions $\mathbf{u} = (c_1(t), \ldots, c_N(t), T(t))$ continuous in some interval $[0, t_1]$ with norm

$$\|\mathbf{u}\| = \max_{0 \leqq t \leqq t_1} |T(t)| + \sum_{i=1}^{N} \max_{0 \leqq t \leqq t_1} |c_i(t)|. \tag{1.8.17}$$

Eqs. (1.8.1), (1.8.2), (1.8.4) are written in integral form as

$$c_i = c_{i0}\, e^{-\frac{t}{\theta}} + \frac{1}{\theta} \int_0^t [c_{if}(\tau) + \theta \sum \nu_{ij} f_j(\mathbf{u}(\tau))]\, e^{-\frac{t-\tau}{\theta}}\, d\tau, \tag{1.8.18}$$

$$\begin{aligned} T = T_0\, e^{-\frac{t}{\theta}} + \frac{1}{\theta} \int_0^t \Big[& T_f(\tau) - \frac{\theta}{C_p} \sum \varDelta H_j f_j(\mathbf{u}(\tau)) \\ & + \frac{\theta}{VC_p} Q(T(\tau)) \Big] e^{-\frac{t-\tau}{\theta}}\, d\tau \end{aligned} \tag{1.8.19}$$

and more succinctly as

$$\mathbf{u}=\mathbf{H}\,\mathbf{u} \tag{1.8.20}$$

where the nonlinear operator $\mathbf{H}$ is clearly completely continuous. We shall show that the vector fields $\mathscr{H}=\mathbf{I}-\mathbf{H}$ and $\mathbf{I}$ are homotopic on any sphere $S_1=\{\mathbf{u}\colon\ \|\mathbf{u}\|=b\}$ surrounding all functions $\mathbf{u}$ subject to the previously derived a priori bounds. As in Section 1.3 we need to observe that any solution of the equation

$$s\mathscr{H}+(1-s)\,\mathbf{I}=0, \qquad 0\leqq s\leqq 1 \tag{1.8.21}$$

or equivalently

$$\frac{dc_i}{dt}=\frac{1}{\theta}(s\,c_{if}-c_i)+s\sum_{j=1}^{R}\nu_{ij}f_j, \tag{1.8.22}$$

$$\frac{dT}{dt}=\frac{1}{\theta}(s\,T_f-T)-\frac{s}{C_p}\sum_{j=1}^{R}\Delta H_j f_j+\frac{s\,Q(T)}{V C_p}, \tag{1.8.23}$$

$$t=0\colon\ c_i=s\,c_{i0}, \quad i=1,\dots,N, \quad T=s\,T_0 \tag{1.8.24}$$

is subject to a priori bounds which are smaller or equal to the a priori bounds derived for the solutions of Eqs. (1.8.1), (1.8.2), (1.8.4). Thus, Eq. (1.8.21) cannot have a solution on the sphere S_1. We have shown now that the vector fields $\mathscr{H}$, $\mathbf{I}$ are homotopic, and hence $\mathscr{H}$ has rotation $+1$ on the sphere S_1. Thus Eqs. (1.8.1), (1.8.2), (1.8.4) have at least one solution in any interval $[0, t_1]$. This result can be stated formally as a theorem.

Theorem 1.7.1. Given continuous bounded functions $\mathbf{u}_f$ and a continuous function $Q(T)$ subject to Eqs. (1.8.12), (1.8.13), the initial value problem (1.8.1), (1.8.2), (1.8.4) has a solution for all $t\geqq 0$.

To insure the uniqueness of the solution some additional restriction, such as a Lipschitz condition, is necessary.

Steady States

As can be seen from Eqs. (1.8.8), (1.8.14), the effect of the initial conditions $\mathbf{u}_0=(c_{10},\dots,c_{N0},T_0)$ quickly disappears, and the state of the system is eventually determined by the input variables $\mathbf{u}_f=(c_{1f},\dots,c_{Nf},T_f)$ alone. More precisely, as $t\to\infty$, the state trajectory rapidly approaches the R dimensional manifold $\Gamma(u_f)\subset E_{N+1}^{+}$ defined by the equations

$$\sum_{i=1}^{N}\gamma_{il}(c_i-c_{if})=0 \qquad i=1,\dots,N-R, \tag{1.8.25}$$

$$T-T_f=\frac{\theta}{C_p V}\,Q(T)-\frac{1}{C_p}\sum_{i=1}^{N}H_i(c_i-c_{if}). \tag{1.8.26}$$

This trajectory lies outside of the manifold $\Gamma(u_f)$ except when $\mathbf{u}_0 \in \Gamma(u_f)$. We shall define the *steady states* of the system as the solutions of the time independent or steady state equations

$$c_i - c_{if} = \theta \sum_{j=1}^{R} \nu_{ij} f_j \qquad i=1, \dots, N, \tag{1.8.27}$$

$$T - T_f = -\frac{\theta}{C_p} \sum_{j=1}^{R} \Delta H_j f_j + \frac{\theta}{V C_p} Q(T). \tag{1.8.28}$$

In the mathematical sense, the steady states are equilibrium points of Eqs. (1.8.1), (1.8.2) but this term will be avoided as having the erroneous connotation of thermodynamic or kinetic equilibrium.

We wish to show the existence of steady states in each manifold $\Gamma(u_f)$ by calculating the rotation of the pertinent vector field. Again, to prevent the occurrence of steady states with negative components we shall follow the procedure and notation of Section 1.4 and define the functions f_j outside of $\Gamma(u_f)$ as

$$\sum_{j=1}^{R} \nu_{ij} f_j(c_1, \dots, c^-, \dots, T) = \sum_{j=1}^{R} \nu_{ij} f_j(c_1, \dots, 0, \dots, T) + (c_{if} - c_i)\, d(\mathbf{u}). \tag{1.8.29}$$

The manifold $\Gamma(u_f)$ can be parametrized by extent variables $\xi_1, \dots, \xi_R$ through the transformation

$$c_i - c_{if} = \sum_{j=1}^{R} \nu_{ij} \xi_j \qquad i=1, \dots, N. \tag{1.8.30}$$

The temperature T can also be expressed in terms of the extents by combining Eqs. (1.8.26), (1.8.30). To simplify matters we shall assume that $T - \theta Q(T)/V C_p$ is strictly increasing and thus insure that Eq. (1.8.26) can be solved uniquely for T. This monotonicity property is satisfied in the common situation in which $Q(T)$ represents the heat exchanged between the reactor fluid and a cooling or heating fluid of constant inlet temperature. Having established a one to one correspondence between $\mathbf{u}$ and $\boldsymbol{\xi}$ we shall denote as usual the image of $\Gamma(u_f)$ in the $\boldsymbol{\xi}$ space by $\tilde{\Gamma}(u_f)$.

We are now ready to prove

Theorem 1.8.2. Each manifold $\Gamma(u_f)$ contains one or more steady states.

Proof. As usual we shall show that the vector fields $\mathbf{I} - \theta \tilde{\mathbf{f}}$ and $\mathbf{I}$ are homotopic on any sphere $S_1 = \{\boldsymbol{\xi}\colon |\boldsymbol{\xi}| = b\}$ surrounding the region $\tilde{\Gamma}(u_f)$. Setting $\mathscr{H}(s) = s(\mathbf{I} - \theta \tilde{\mathbf{f}}) + (1-s)\mathbf{I}$ we verify that $\mathscr{H}(0) = \mathbf{I}$, $\mathscr{H}(1) = \mathbf{I} - \theta \tilde{\mathbf{f}}$ and it remains to show that $\mathscr{H}(s)$ has no null points on S_1 for any

$s \in [0, 1]$. If $\xi \in S_1$ is a null point of $\mathscr{H}(s)$ then

$$\xi_j = s\,\theta\,\tilde{f}_j \qquad j = 1, \ldots, R. \tag{1.8.31}$$

On the other hand, since $S_1 \subset E_R - \tilde{\Gamma}(u_f)$ each point $\xi \in S_1$ corresponds to a state $\mathbf{u}$ with one or more negative components. If that component is $c_i < 0$ we can introduce Eq.(1.8.31) in Eq.(1.8.29) and arrive at the contradiction

$$0 \leqq \sum_{j=1}^{R} \nu_{ij} f_j(c_1, \ldots, 0, \ldots, T) = -(c_{if} - c_i)\left(d(\mathbf{u}) + \frac{1}{s\,\theta}\right) < 0. \tag{1.8.32}$$

If the negative component is T we obtain from Eqs.(1.8.29), (1.8.31) $c_k = c_{kf}$, $k = 1, \ldots, N$ so that Eq.(1.8.26) reads

$$T - \frac{\theta}{V C_p} Q(T) = T_f. \tag{1.8.33}$$

This relation contradicts the monotonically increasing character of the function $T - \theta Q(T)/V C_p$, which evaluated at $T = 0$ gives $Q(0) > 0$. The vector fields $\mathbf{I} - \theta\tilde{\mathbf{f}}$ and $\mathbf{I}$ have now been shown to be homotopic and hence they have the same rotation, $+1$. This implies that Eqs.(1.8.27), (1.8.28) have one or more solutions in $\Gamma(u_f)$.

1.9. Uniqueness and Stability of Steady States in Open Systems

We shall turn now to the stirred tank reactor, which has been already studied in Section 1.8 relative to the existence of steady states. In this section we shall show that the steady states are odd in number, $2m+1$, among which m at least are unstable. Moreover, for values of the parameter θ outside a certain range, the steady state is unique and stable. This will be the extent of our occupation with the stability problem. For standard work on the dynamics and stability of stirred tank reactors, including detailed phase space analyses, see the textbook of Aris [3], the work of Aris and Amundson [6, 7], and Luus and Lapidus [32].

A steady state $\boldsymbol{\xi}_s \in \tilde{\Gamma}(u_f)$ is a solution of

$$\boldsymbol{\xi} = \theta\,\tilde{\mathbf{f}}(\boldsymbol{\xi}) \tag{1.9.1}$$

where the extents ξ are defined by

$$c_i - c_{if} = \sum_{j=1}^{R} \nu_{ij}\,\xi_j. \tag{1.9.2}$$

The index of $\boldsymbol{\xi}_s$ is determined from the eigenvalue problem

$$\boldsymbol{\phi} = \lambda\,\theta\,\mathbf{A}(\boldsymbol{\xi}_s)\,\boldsymbol{\phi} \tag{1.9.3}$$

where $\mathbf{A}(\xi)$ is the $R \times R$ matrix of Eq. (1.7.5). According to Theorem A.5, the index of ξ_s is equal to $(-1)^\beta$, where β is the sum of multiplicities of the eigenvalues of the matrix $\mathbf{A}(\xi_s)$ in the interval $(1/\theta, \infty)$. On the other hand, the vector field $\mathbf{I} - \theta \tilde{\mathbf{f}}$ has rotation $+1$ over any sphere surrounding $\tilde{\Gamma}(u_f)$. Therefore, the number of steady states is odd, $2m+1$, with m having index -1 and $m+1$ having index $+1$.

To understand the exceptional case of $1/\theta$ being an eigenvalue of the matrix $\mathbf{A}(\xi_s)$ we should note that Eqs. (1.9.1) define implicitly a transformation from the variables $\theta, c_{1f}, \ldots, c_{Nf}, T_f$ to the variables $\xi_1, \ldots, \xi_R$. Thus, when the Jacobian matrix $\mathbf{A}(\xi_s) - \frac{1}{\theta}\mathbf{I}$ is singular, Eqs. (1.9.1) cannot be solved uniquely in the neighborhood of $(\theta, \mathbf{u}_f)$, ξ_s. In many cases, a small perturbation of the variables $\theta, \mathbf{u}_f$ will either annihilate ξ_s or split it in two neighboring steady states with indices $+1$, -1, respectively. This matter, however, is not entirely understood and deserves further investigation.

In the two limiting cases of small θ and large θ the index of a steady state is always $+1$ and the steady state is unique. The uniqueness property for small θ is a standard property of nonlinear equations of the type of Eq. (1.9.1). The uniqueness for large θ, however, is a rather remarkable characteristic of chemical reaction systems. We shall consider first the case of small θ and define quantities

$$B(\xi, \rho) = \sum_{i,j=1}^{R} A_{ij}(\xi)\, \rho_i\, \rho_j, \tag{1.9.4}$$

$$b(\xi) = \max_{|\rho|=1} B(\xi, \rho), \tag{1.9.5}$$

$$b_M = \max_{\xi \in \tilde{\Gamma}(u_f)} b(\xi). \tag{1.9.6}$$

The real parts of the eigenvalues of the matrix $\mathbf{A}(\xi_s)$ are clearly smaller than or equal to b_M, therefore if

$$\theta\, b_M < 1 \tag{1.9.7}$$

none of these eigenvalues lies in the interval $(1/\theta, \infty)$ and the index of ξ_s is $+1$. It is also easy to verify that when $\theta\, b_M < 1$, Eq. (1.9.1) is a contraction mapping and can be solved by the method of successive approximations. The dimensionless parameter $\theta\, b_M$ has approximately the dependence

$$\theta\, b_M \sim \frac{V|f_i(\mathbf{u}_f)|}{w\, c_{if}} \tag{1.9.8}$$

where the numerator and denominator represent a reaction rate and a flow rate respectively. Small $\theta\, b_M$ signifies a small ratio of reaction to flow rate so that $\mathbf{u}$ is close to $\mathbf{u}_f$ and there is no possibility for multiple steady states and instabilities. The regime of small θb_M may be called the reaction limited regime.

The case of large residence times θ is subtler. Let us start by recalling that an equilibrium state ξ^* is a solution of Eq. (1.7.1) and a steady state ξ_s is a solution of Eq. (1.9.1) and depends on θ and $\mathbf{u}_f$. Suppose now that the equilibrium state $\xi^* \in \tilde{\Gamma}(u_f)$ is unique and stable. The steady state in $\tilde{\Gamma}(u_f)$ need not be unique in all cases, for it depends on the interaction of reaction rates and flow rate. We wish to show that ξ_s is unique for sufficiently large θ. To this end we set

$$\xi_M = \max_{\xi \in \tilde{\Gamma}(u_f)} |\xi| \tag{1.9.9}$$

and obtain from Eq. (1.9.1)

$$|\tilde{\mathbf{f}}(\xi_s)| \leqq \frac{\xi_M}{\theta} \tag{1.9.10}$$

so that for large θ, $\tilde{\mathbf{f}}(\xi_s) \to 0$. Since ξ^* is stable, all the eigenvalues of the matrix $\mathbf{A}(\xi^*)$ have negative real parts. We can choose then a $\delta > 0$ such that all eigenvalue of the matrix $\mathbf{A}(\xi)$ have negative real parts whenever $|\xi - \xi^*| < \delta$. That this choice of δ is possible can be seen in KATO [21], Th. 5.1. On the other hand, Eq. (1.9.10) implies that given any $\delta > 0$ there exists a θ_2 such that $|\xi_s(\theta) - \xi^*| < \delta$ for all $\theta \geqq \theta_2$. Otherwise a sequence $\{\xi_s(\theta_n)\}$ could be chosen with the properties

$$\lim_{n \to \infty} \tilde{\mathbf{f}}(\xi_s(\theta_n)) = 0, \tag{1.9.11}$$

$$|\xi_s(\theta_n) - \xi^*| > \delta. \tag{1.9.12}$$

Since $\tilde{\Gamma}(\mathbf{u}_f)$ is closed, the sequence $\{\xi_s(\theta_n)\}$ has an accumulation point $\bar{\xi} \neq \xi^*$ with $\tilde{\mathbf{f}}(\bar{\xi}) = 0$, a result contradicting the uniqueness of the equilibrium state ξ^*. Therefore, $\theta > \theta_2$ insures that the eigenvalues of the matrix $\mathbf{A}(\xi_s(\theta))$ have negative real parts and that $\xi_s(\theta)$ has index $+1$.

An intuitive understanding of the situation may be obtained from Eq. (1.9.10) by performing the order of magnitude estimates

$$\tilde{\mathbf{f}}(\xi_s) \sim \tilde{\mathbf{f}}(\xi^*) + \mathbf{A}(\xi^*)(\xi_s - \xi^*) = \mathbf{A}(\xi^*)(\xi_s - \xi^*), \tag{1.9.13}$$

$$b_M |\xi_s - \xi^*| \leqq \frac{\xi_M}{\theta}. \tag{1.9.14}$$

Thus when $b_M \theta$ is sufficiently large, the steady state ξ_s lies close to ξ^* and there is no opportunity for multiple steady states and instabilities. The regime of large $b_M \theta$ may be called the flow (or transport) limited regime.

The foregoing results will be stated as a theorem.

Theorem 1.9.1. Let $\xi_s^{(1)}, \ldots, \xi_s^{(n)}$ be the steady states in a given manifold $\tilde{\Gamma}(u_f)$. If none of the matrices $\mathbf{A}(\xi_s^{(i)})$ has the number $1/\theta$ as an eigenvalue, the number of steady states is odd, $n=2m+1$. Among these steady states m have index -1 and $m+1$ have index $+1$. For sufficiently small θ Eq. (1.9.1) is a contraction mapping and the steady state is unique. If the equilibrium state $\xi^* \in \tilde{\Gamma}(u_f)$ is unique and stable, the steady state ξ_s is unique for sufficiently large values of θ.

Stability

In open systems it is natural to consider perturbations in the initial state and the feed state, which suggests the following definition of stability. A steady state $\mathbf{u}_s \in \Gamma(u_f)$ will be called stable if given any $\varepsilon>0$ there exists a $\delta>0$ such that $|\mathbf{u}_0-\mathbf{u}_s|<\delta$, $|\mathbf{u}'_f-\mathbf{u}_f|<\delta$ imply that the solution $\mathbf{u}(t)$ of the equations

$$\frac{dc_i}{dt}=\frac{1}{\theta}(c'_{if}-c_i)+\sum_{j=1}^{R}\nu_{ij}f_j \qquad i=1,\ldots,N, \tag{1.9.15}$$

$$\frac{dT}{dt}=\frac{1}{\theta}(T'_f-T)-\frac{1}{C_p}\sum_{j=1}^{R}\Delta H_j f_j+\frac{Q(T)}{C_p} \tag{1.9.16}$$

$$t=0:\ \mathbf{u}=\mathbf{u}_0 \tag{1.9.17}$$

satisfies

$$|\mathbf{u}(t)-\mathbf{u}_s|<\varepsilon, \qquad t\geqq 0. \tag{1.9.18}$$

As in the previous section, it can be shown that the stability of the steady state $\mathbf{u}_s \in \Gamma(u_f)$ under the perturbations $\mathbf{u}_0$, $\mathbf{u}'_f$ is equivalent to the stability of the steady state $\mathbf{u}'_s \in \Gamma(u'_f)$ under the perturbation $\mathbf{u}_0$. To study the latter problem let us linearize Eqs. (1.9.15)–(1.9.18) about $\mathbf{u}'_s$

$$\frac{dv_i}{dt}=-\frac{1}{\theta}v_i+\sum_{j=1}^{R}\nu_{ij}\left[\sum_{k=1}^{N+1}\left(\frac{\partial f_j}{\partial u_k}\right)' v_k\right] \qquad i=1,\ldots,N, \tag{1.9.19}$$

$$\frac{dv_{N+1}}{dt}=-\frac{1}{\theta}v_{N+1}-\frac{1}{C_p}\sum_{j=1}^{R}\Delta H_j\left[\sum_{k=1}^{N+1}\left(\frac{\partial f_j}{\partial u_k}\right)' v_k\right]+\left(\frac{dQ}{dT}\right)' v_{N+1} \tag{1.9.20}$$

where the prime denotes that the derivatives are evaluated at $\mathbf{u}'_s$.

We wish to utilize the fact that as $t\to\infty$ the state of the system approaches the manifold $\Gamma(u'_f)$. This property, which has been shown in Section 1.8, depends in no way on the stability characteristics of $\mathbf{u}'_s$. Thus, as pointed out by Aris [3], the stability of $\mathbf{u}'_s$ depends on the convergence characteristics of the trajectories which lie entirely in the manifold

$\Gamma(u_f')$. To show this we define new variables $\phi_1, \dots, \phi_R, \psi_1, \dots, \psi_{N-R+1}$ by

$$v_i = \sum_{j=1}^{R} \nu_{ij} \phi_j \qquad i=1, \dots, R, \tag{1.9.21}$$

$$\psi_l = \sum_{i=1}^{N} \gamma_{il} v_i \qquad l=1, \dots, N-R, \tag{1.9.22}$$

$$\psi_{N-R+1} = \frac{1}{C_p} \sum_{i=1}^{N} H_i v_i + \left[1 - \frac{\theta}{VC_p} \left(\frac{dQ}{dT}\right)'\right] v_{N-R+1}. \tag{1.9.23}$$

By introducing Eq.(1.9.21) in Eq.(1.9.19) there is obtained

$$\frac{d\phi_j}{dt} = -\frac{1}{\theta} \phi_j + \sum_{\rho=1}^{R} \left[\sum_{k=1}^{N+1} \left(\frac{\partial f_j}{\partial u_k}\right)' \nu_{k\rho}\right] \phi_\rho \qquad j=1, \dots, R. \tag{1.9.24}$$

By differentiating Eqs.(1.9.22), (1.9.23) and utilizing the expressions (1.9.19), (1.9.20) for the derivatives dv_i/dt there is obtained

$$\frac{d\psi_l}{dt} = -\frac{1}{\theta} \psi_l \qquad l=1, \dots, N-R+1. \tag{1.9.25}$$

It is clear that the variables $\phi_1, \dots, \phi_R, \psi_1, \dots, \psi_{N-R+1}$ are linearly independent and hence the systems (1.9.19), (1.9.20) and (1.9.24), (1.9.25) are equivalent. The stability of the original problem reduces, therefore, to the stability of the system (1.9.24).

To cast Eqs.(1.9.24) in a more familiar form we define extents

$$c_i - c_{if}' = \sum_{j=1}^{R} \nu_{ij} \xi_j' \qquad i=1, \dots, N \tag{1.9.26}$$

and observe from Eq.(1.7.5) and the definition of $\tilde{f}_j$

$$\tilde{f}_j(\xi) = f_j(\mathbf{u}(\xi'; \mathbf{u}_f')) \tag{1.9.27}$$

that

$$A_{jk}(\xi_s') = \frac{\partial \tilde{f}_j}{\partial \xi_k} = \sum_{l=1}^{N+1} \left(\frac{\partial f_j}{\partial u_l}\right)' \nu_{lk} \tag{1.9.28}$$

where ξ_s' is the image of $\mathbf{u}_s'$ in $\tilde{\Gamma}(u_f')$. Eqs.(1.9.24) can, therefore, be written as

$$\frac{d\boldsymbol{\phi}}{dt} = -\frac{1}{\theta} \boldsymbol{\phi} + \mathbf{A}(\xi_s')\boldsymbol{\phi}. \tag{1.9.29}$$

If the index of ξ_s' is -1, the matrix $\mathbf{A}(\xi_s')$ has one at least eigenvalue in $(1/\theta, \infty)$ so that by Eq.(1.9.29) the steady state ξ_s' is unstable. On the other hand, if the index of ξ_s' is $+1$ this steady state is not necessarily stable.

It has been shown earlier in this section that when θ is sufficiently small or sufficiently large none of the eigenvalues of the matrix $\mathbf{A}(\boldsymbol{\xi}_s(\theta))$ has real parts in $(1/\theta, \infty)$ and the steady state $\boldsymbol{\xi}_s$ is unique. Under the same conditions, no eigenvalue of Eq. (1.9.29) has positive real parts so that the steady state $\boldsymbol{\xi}_s$ is stable.

Theorem 1.9.2. Under the conditions of Theorem 1.9.1, at least m of the $2m+1$ steady states of a given manifold are unstable. The steady state is unique and stable provided θ is either sufficiently small or sufficiently large.

Example 1.9.1. The concepts of the present section can be illustrated by the simple example of a single irreversible exothermic reaction in a stirred tank reactor with no heat exchange ($Q=0$).

$$\mathcal{M}_2-\mathcal{M}_1=0, \tag{1.9.30}$$

$$f=k\,c_1\,e^{-\frac{E}{R_g T}}. \tag{1.9.31}$$

This system is described by the time dependent equations

$$\frac{dc_1}{dt}=\frac{1}{\theta}(c_{1f}-c_1)-k\,c_1\,e^{-\frac{E}{R_g T}}, \tag{1.9.32}$$

$$\frac{dT}{dt}=\frac{1}{\theta}(T_f-T)+\frac{|\Delta H|}{C_p}\,k\,c_1\,e^{-\frac{E}{R_g T}} \tag{1.9.33}$$

and the steady state equations

$$c_{1f}-c_1=\theta\,k\,c_1\,e^{-\frac{E}{R_g T}}, \tag{1.9.34}$$

$$T_f-T=-\frac{\theta|\Delta H|}{C_p}\,k\,c_1\,e^{-\frac{E}{R_g T}}. \tag{1.9.35}$$

By introducing the extent ξ

$$c_1-c_{1f}=-\xi, \tag{1.9.36}$$

$$T-T_f=\frac{|\Delta H|}{C_p}\,\xi, \tag{1.9.37}$$

we can reduce Eqs. (1.9.32), (1.9.33) to

$$\xi=\theta\,\tilde{f}(\xi) \tag{1.9.38}$$

where

$$\tilde{f}(\xi)=k(c_{1f}-\xi)\exp\left[-\frac{C_p E}{R_g(C_p T_f+\xi|\Delta H|)}\right]. \tag{1.9.39}$$

The invariant manifold $\tilde{\Gamma}(u_f)$ is simply the interval $[0, c_{1f}]$.

A graphical solution of Eq. (1.9.38) is shown schematically in Fig. 1.9.1, where it has been assumed that $c_{1f} E |\Delta H| > C_p R_g T_f^2$. When θ is smaller than θ_2 or larger than θ_4 the steady state (e.g. ξ_1, ξ_9) is unique and stable. When $\theta = \theta_4$ or $\theta = \theta_2$, $d\tilde{f}/d\xi = 1/\theta$ and the index of the points ξ_4, ξ_6 is

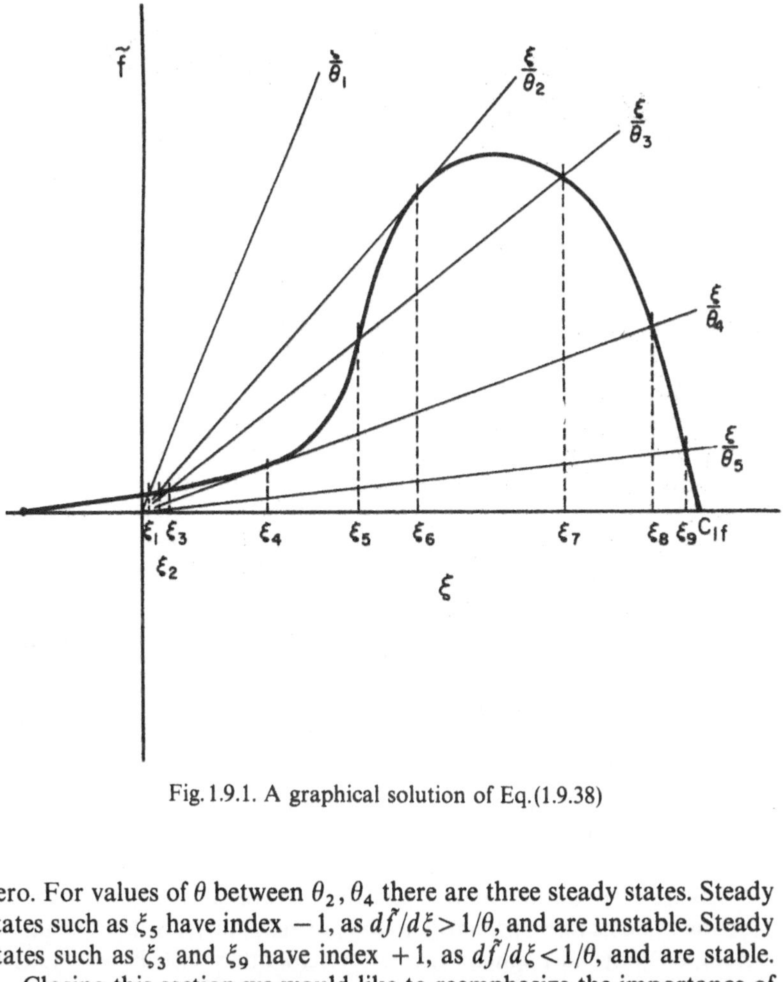

Fig. 1.9.1. A graphical solution of Eq. (1.9.38)

zero. For values of θ between θ_2, θ_4 there are three steady states. Steady states such as ξ_5 have index -1, as $d\tilde{f}/d\xi > 1/\theta$, and are unstable. Steady states such as ξ_3 and ξ_9 have index $+1$, as $d\tilde{f}/d\xi < 1/\theta$, and are stable.

Closing this section we would like to reemphasize the importance of the dimensionless parameter $b_M \theta$ or $V|f_i(u_f)|/w c_{if}$. Multiple steady states and instabilities are possible only when $b_M \theta$ is of the order of unity, i.e., when the flow rate and the reaction rate balance each other. If either of the two rates predominates the steady state is unique and stable.

Chapter 2

Distributed Chemical Reaction Systems

2.1. Physical Description and Formulation of the Conservation Differential Equations

A distributed chemical reaction system occupying a region V with a bounding surface S is completely characterized by the state variables $\mathbf{u}(r)=(c_1(r), \ldots, c_N(r), T(r))$ and the mean velocity $\mathbf{v}(r)$ given at each point $r \in V$. Diffusive fluxes of chemical species and thermal energy exist at each point of the system due to gradients in the concentrations and the temperature. These gradients are maintained by means of the chemical reactions, acting as sources or sinks of chemical species and thermal energy. System and surroundings exchange chemical species and thermal energy through the bounding surface S.

As examples of distributed chemical reaction systems we can mention the porous catalyst pellet and the membrane of a cell, which both consist of an inhomogeneous supporting phase, solid or otherwise, with well defined boundaries. Fluids diffuse through the boundaries in the supporting phase where they react, and the products diffuse out. Chemical reactions and transport phenomena have interesting interactions in distributed systems. For example, the product distribution of a catalytic reaction depends on the relative magnitude of the diffusivities of the gases in the catalyst. Similarly, the presence of a chemical reaction may increase the transport of a chemical species through a cell membrane.

In the systems under consideration the supporting phase either does not undergo any chemical change or it changes at a rate much slower than the rate of the other reactions, as in the case of catalyst poisoning. Reactions such as the formation of a metal oxide whereby the solid itself is subject to the main chemical change will not be considered in this work.

Our treatment of the mathematical structure of distributed chemical reaction systems refers primarily to the porous catalyst pellet. The theory developed may be applicable, however, to other distributed systems. The catalyst pellet has been extensively investigated, experimentally and theoretically, due to its practical importance. The catalytic reactors, so common in the chemical and petrochemical industry, are devices contacting a fluid with a fixed or fluidized bed of catalyst pellets. Thus, an understanding of the properties of a single pellet is essential to the understanding of the reactor's operation. Catalytic reactions are fast and often release large amounts of thermal energy. This rather violent behavior manifests itself with the possibility of multiple steady states and instabilities, phenomena which have fascinated the theoretical engineer.

A porous catalyst pellet usually consists of a close packing of small porous catalyst particles of size 10–1000 Å. The gases diffuse in the empty space between the particles as well as in the microscopic pores of the particles. The most important physical property of such a structure is the area per unit mass of the pellet, denoted by S_p. The values of S_p are in the range of $1-1000\,\mathrm{m}^2/\mathrm{g}$. The mass per unit of total volume ρ_p, and the void fraction ε_p are also essential for the physical characterization of the catalyst. In many cases the porous pellets are impregnated with a solution of a metal salt and subsequently heated or treated chemically. The microscopic crystals of the metal or metal oxide deposited on the porous surface by this process give the catalyst its chemical reactivity.

Before starting with the analysis of the mathematical models describing distributed chemical reaction systems we shall give a brief physical discussion concerning the physical basis and the limitations of the expressions for the reaction rates and the fluxes in porous catalyst pellets.

For an introduction to the chemical engineering aspects of chemical reactions and transport in porous catalysts we refer to the comprehensive textbook of PETERSEN [34]. This textbook also provides an access to the literature on the subject.

Heterogeneous Catalytic Reactions

A simple picture of a heterogeneous catalytic reaction can be given as follows. Molecules from the fluid phase are adsorbed on *active sites* of the catalytic surface with which they form complexes. The site-molecule complexes react to products still adsorbed on the surface. Finally, the products desorb to the fluid phase regenerating the catalyst sites. An example taken from PETERSEN [34]* will introduce some important ideas. Consider the isomerization reaction:

$$-\mathcal{M}_1+\mathcal{M}_2=0 \tag{2.1.1}$$

proceeding according to the mechanism:

$$-\mathcal{M}_1-\mathcal{X}_1+\mathcal{X}_2=0, \tag{2.1.2}$$

$$-\mathcal{X}_2+\mathcal{X}_3=0, \tag{2.1.3}$$

$$-\mathcal{X}_3+\mathcal{X}_1+\mathcal{M}_2=0. \tag{2.1.4}$$

According to Eq. (2.1.2), $\mathcal{M}_1$ is adsorbed on an active site $\mathcal{X}_1$ forming the complex $\mathcal{X}_2$. This complex reacts according to Eq. (2.1.3) to form a

* Reprinted by permission of Prentice-Hall, Inc.

complex $\mathscr{X}_3$ which decomposes according to Eq.(2.1.4) with the desorption of the product $\mathscr{M}_2$ and the regeneration of the active site $\mathscr{X}_1$. We let c_1, c_2 be the concentrations of $\mathscr{M}_1, \mathscr{M}_2$ in the gas phase right at the surface and x_1, x_2, x_3 the concentrations of $\mathscr{X}_1, \mathscr{X}_2, \mathscr{X}_3$ on the solid surface. While c_1, c_2 have units of moles per unit volume, x_1, x_2, x_3 have units of moles per unit surface. We shall assume that the reactions (2.1.2), (2.1.3), (2.1.4) have the rates

$$f_1' = k_1^+ c_1 x_1 - k_1^- x_2, \tag{2.1.5}$$

$$f_2' = k_2^+ x_2 - k_2^- x_3, \tag{2.1.6}$$

$$f_3' = k_3^+ x_3 - k_3^- c_2 x_1 \tag{2.1.7}$$

where $k_1^+, \ldots, k_3^-$ are functions of temperature only. The conservation equations are

$$\frac{dx_1}{dt} = -f_1' + f_3' = -(k_1^+ c_1 + k_3^- c_2) x_1 + k_1^- x_2 + k_3^- x_3, \tag{2.1.8}$$

$$\frac{dx_2}{dt} = f_1' - f_2' = k_1^+ c_1 x_1 - (k_1^- + k_2^+) x_2 + k_2^- x_3, \tag{2.1.9}$$

$$\frac{dx_3}{dt} = f_2' - f_3' = k_3^- c_2 x_1 + k_2^+ x_2 - (k_2^- + k_3^+) x_3, \tag{2.1.10}$$

$$\frac{dc_1}{dt} = -f_1' = -k_1^+ c_1 x_1 + k_1^+ x_2, \tag{2.1.11}$$

$$\frac{dc_2}{dt} = f_3' = -k_3^- c_2 x_1 + k_3^+ x_3. \tag{2.1.12}$$

By adding Eqs.(2.1.8)–(2.1.10) it is verified that the total number of sites is constant

$$x_1 + x_2 + x_3 = x_0. \tag{2.1.13}$$

At this point one often applies the steady state approximation [34] for the species $\mathscr{X}_1, \mathscr{X}_2, \mathscr{X}_3$ by setting

$$\frac{dx_1}{dt} = -(k_1^+ c_1 + k_3^- c_2) x_1 + k_1^- x_2 + k_3^- x_3 = 0, \tag{2.1.14}$$

$$\frac{dx_2}{dt} = k_1^+ c_1 x_1 - (k_1^- + k_2^+) x_2 + k_2^- x_3 = 0, \tag{2.1.15}$$

$$\frac{dx_3}{dt} = k_3^- c_2 x_1 + k_2^+ x_2 - (k_2^- + k_3^+) x_3 = 0 \tag{2.1.16}$$

and solving to obtain expressions for x_1, x_2, x_3 in terms of c_1, c_2, x_0. These expressions can be introduced in Eqs. (2.1.11), (2.1.12) to obtain the rates in terms of c_1, c_2 alone. The resulting formulas are cumbersome but further simplification is often possible under certain conditions. Suppose for example that the reaction (2.1.3) is much slower than the adsorption and desorption. Quantitatively this means that the terms $k_1^+ c_1 x_1, k_1^- x_2, k_3^+ x_3, k_3^- x_1 c_2$ are all much larger than $k_2^+ x_2, k_2^- x_3$. In such a case the reaction (2.1.3) is said to be *rate controlling*. Eqs. (2.1.14) to (2.1.16) can then be simplified to

$$k_1^+ c_1 x_1 - k_1^- x_2 = 0, \tag{2.1.17}$$

$$k_3^- c_2 x_1 - k_3^+ x_3 = 0 \tag{2.1.18}$$

which means that the adsorption and desorption steps (2.1.2), (2.1.4) are very near equilibrium while the reaction (2.1.3) is far from equilibrium in such a way that all three rates equal. The solution of Eqs. (2.1.17), (2.1.18), (2.1.13) is

$$x_1 = \frac{x_0}{\alpha}, \tag{2.1.19}$$

$$x_2 = \frac{k_1^+ x_0}{k_1^- \alpha} c_1, \tag{2.1.20}$$

$$x_3 = \frac{k_3^- x_0}{k_3^+ \alpha} c_2 \tag{2.1.21}$$

where

$$\alpha = 1 + \frac{k_1^+}{k_1^-} c_1 + \frac{k_3^-}{k_3^+} c_2. \tag{2.1.22}$$

Now, we have used $f_1' = 0, f_3' = 0$ to calculate the concentrations x_1, x_2, x_3. Actually, f_1', f_3' are small but not zero. Their value can be obtained from $f_1' = f_3' = f_2'$ where f_2' in turn is obtained by introducing Eqs. (2.1.20), (2.1.21) in Eq. (2.1.6)

$$f_1' = f_2' = f_3' = f' = \frac{x_0}{\alpha} \left[\frac{k_1^+ k_2^+}{k_1^-} c_1 - \frac{k_2^- k_3^-}{k_3^+} c_2 \right] \tag{2.1.23}$$

where the common rate f' is also the rate of the overall reaction (2.1.1). In the limiting case $(k_1^+/k_1^-) c_1$, $(k_3^-/k_3^+) c_2 \ll 1$, we obtain the first order kinetics

$$f' = x_0 \left[\frac{k_1^+ k_2^+}{k_1^-} c_1 - \frac{k_2^- k_3^-}{k_3^+} c_2 \right]. \tag{2.1.24}$$

If in addition $(k_2^- k_3^-/k_3^+) c_2 \ll (k_1^+ k_2^+/k_1^-) c_1$ we obtain the irreversible first order kinetics

$$f' = \frac{x_0 k_1^+ k_2^+}{k_1^-} c_1. \tag{2.1.25}$$

The foregoing derivations show a possible mechanistic explanation of some of the simple kinetic expressions given in the literature.

Now x_0, x_1, x_2, x_3 have units of moles per unit area and hence f' has units of moles per unit area per unit time. The equivalent rate

$$f=\rho_p S_p f' \tag{2.1.26}$$

given in moles per unit volume per unit time, is the correct source term to be used with the conservation equations.

Diffusion in Porous Media

A precise description of diffusion in porous media has not been achieved so far. Current efforts aim at developing expressions such as $N_i = D_i \operatorname{grad} c_i$ where c_i is the concentration of the ith species per unit of void or intersticial volume (total volume minus solid volume), N_i is the flux of the ith species defined relative to total (solid and void) area, and D_i is an *effective diffusivity*, generally a function of $c_1, \ldots, c_N, T$. In the case of a very fine pore structure, $\lambda/r_p \gg 1$ (where λ is the mean free path in the gas and r_p the pore diameter), the gas molecules collide with the surface of the pores but not with each other and each species diffuses independently. This is called Knudsen diffusion and is described by

$$N_i = -g\, D_i^K \operatorname{grad} c_i \tag{2.1.27}$$

where the Knudsen diffusivity D_i^K is given by

$$D_i^K = \frac{4\, r_p}{3} \left(\frac{2\, R_g\, T}{\pi\, M_i} \right)^{\frac{1}{2}}. \tag{2.1.28}$$

The geometric factor g, accounting for the structure of the porous medium, should be obtained experimentally.

Diffusion in the continuum regime $(\lambda/r_p) \ll 1$ can be described by the usual molecular diffusion coefficient. However, the treatment of multi-component mixtures is very tedious except when they can be approximately considered as binary. Diffusion in the transition regime $(\lambda/r_p \sim 1)$ has been described by effective diffusivities which are functions of the molecular and the Knudsen diffusivity [34].

A common difficulty in all the aforementioned cases is the irregular geometry of the porous medium. In addition, a precise analysis will have to consider that the diffusion in the pores is modified by surface diffusion and surface reactions. A completely phenomenological approach based on irreversible thermodynamics would give theoretically consistent transport expressions but would be too complicated for experimental determination of the transport coefficients and for the solution of the conservation equations.

In view of the difficulties in employing a precise description of diffusion in porous catalysts the simple expression

$$N_i = -D_i \operatorname{grad} c_i \tag{2.1.29}$$

with a constant D_i has found widespread use. The error introduced by this formula is often smaller than the uncertainty involved in the experimental determination of the reaction rates.

Heat conduction in porous catalysts is well described by the expression

$$q = -k_c \operatorname{grad} T \tag{2.1.30}$$

where k_c is an *effective* thermal conductivity such that q is the heat flux per unit total area. Since the thermal conductivity of gases is much smaller than that of solids, k_c depends on the properties of the porous medium and is practically independent of the gases present.

The Conservation Equations

We shall limit our attention to spherical regions and occasionally to planar regions. With the geometry in its simplest form, the analysis will be directed at the nonlinearities induced by the chemical reactions. Most of the results could be extended to regions of arbitrary shape. Such extensions, however, would be very tedious requiring the use of more complex function spaces and the consideration of questions of boundary smoothness.

By using Eqs. (2.1.26), (2.1.29) the conservation equations for a spherical region may be written as

$$\varepsilon_p \frac{\partial c_i}{\partial t} = \frac{1}{r^2} \frac{\partial}{\partial r}\left(r^2 D_i \frac{\partial c_i}{\partial r}\right) + \sum_{j=1}^{R} \nu_{ij} f_j, \tag{2.1.31}$$

$$\rho_p c_p \frac{\partial T}{\partial t} = \frac{1}{r^2} \frac{\partial}{\partial r}\left(r^2 k_c \frac{\partial T}{\partial r}\right) - \sum_{j=1}^{R} \Delta H_j f_j \tag{2.1.32}$$

where

$$\Delta H_j = \sum_{i=1}^{N} \nu_{ij} H_i. \tag{2.1.33}$$

Note again that the concentrations c_i are given in moles per unit void volume, while the "pseudohomogeneous" rates f_j, defined by Eq. (2.1.26), are given in moles per unit total volume per unit time. Eq. (2.1.32) is a simplified form of the energy equation. It is a heat conduction equation with a chemical reaction source term and partly neglects the variation with temperature of the enthalpies.

The large thermal conductivity of the solid catalyst tends to equalize the temperature so that variations in the quantities T, $|\Delta H_j|$, k_c are much smaller than these quantities themselves. Thus, ΔH_j and k_c are approximately constant in the catalyst pellet. The diffusivities $D_1, \ldots, D_N$ also depend moderately on the temperature and can be taken as constants except when they vary significantly with composition. On the other hand, the reaction rates $f_1, \ldots, f_R$ depend very strongly on the temperature, whence the necessity of considering Eq. (2.1.32).

The boundary and initial conditions associated with Eqs. (2.1.31) and (2.1.32) will be formulated in the following sections as it becomes necessary.

The time independent form of Eqs. (2.1.32), (2.1.33) is

$$\frac{1}{r^2}\frac{d}{dr}\left(r^2 D_i \frac{dc_i}{dr}\right) = -\sum_{j=1}^{R} \nu_{ij} f_j, \tag{2.1.34}$$

$$\frac{1}{r^2}\frac{d}{dr}\left(r^2 k_c \frac{dT}{dr}\right) = \sum_{j=1}^{R} \Delta H_j f_j. \tag{2.1.35}$$

The solutions of Eqs. (2.1.34), (2.1.35) with appropriate boundary conditions such as Eqs. (2.4.3), (2.4.4) will be called the *steady states* of the system. Various properties of the steady states, such as the invariant manifolds and a priori bounds, the existence and uniqueness of solutions, the asymptotic behavior, and the stability will be treated in the sections that follow. There is a strong similarity in the properties of the uniform open systems investigated in Sections 1.8, 1.9 and the distributed systems to be studied now. In both types of systems the interplay between reaction and transport rates (or flow rates) creates the possibility of multiple steady states for certain types of reaction kinetics. Furthermore, the conditions for uniqueness and stability of the steady state have a common mathematical and physical basis.

As with uniform systems we shall focus attention to nonlinear reaction kinetics. The important special case of linear kinetics has been treated in detail by WEI [41].

2.2. One Reaction. Invariant Manifolds and A Priori Bounds

Distributed systems with a single chemical reaction present the simplest problem for analysis due to the fact that at steady state they can be described by a single state variable. In particular, when the geometry is planar (a region bounded by two infinite parallel planes), the steady equations accept a closed form solution which can be used to obtain results about uniqueness and asymptotic behavior. In addition to the availability of a closed form, the single reaction is a convenient starting point for developing concepts and deriving results to be generalized later to systems with many reactions. For it should be stated from the beginning that the basic results about existence, uniqueness, asymptotic behavior, and stability are independent of the number of chemical reactions.

The time independent equations describing a system with the single reaction

$$\sum_{i=1}^{N} \nu_i \mathcal{M}_i = 0 \tag{2.2.1}$$

are

$$\frac{1}{r^2}\frac{d}{dr}\left(r^2 D_i \frac{dc_i}{dr}\right) = -\nu_i f \qquad i=1,\ldots,N, \tag{2.2.2}$$

$$\frac{1}{r^2}\frac{d}{dr}\left(r^2 k_c \frac{dT}{dr}\right) = \Delta H f. \tag{2.2.3}$$

Initially we shall consider the boundary conditions

$$r=0:\ \frac{dc_i}{dr}=0, \qquad \frac{dT}{dr}=0, \tag{2.2.4}$$

$$r=r_0:\ c_i=c_{i0}, \qquad T=T_0. \tag{2.2.5}$$

The boundary condition (2.2.4) results from the symmetry of the system-spherical region with uniform surface conditions. The boundary condition (2.2.5) can be physically realized if the resistance for heat and mass transport between the spherical region and the surrounding fluid is much smaller than the resistance for transport in the interior of the region. Then the surface of the region is kept at very nearly the same state as that of the surrounding medium.

By eliminating f from Eqs. (2.2.2), (2.2.3) there is obtained

$$\frac{1}{r^2}\frac{d}{dr}\left[r^2\left(\nu_j D_i \frac{dc_i}{dr} - \nu_i D_j \frac{dc_j}{dr}\right)\right]=0, \tag{2.2.6}$$

$$\frac{1}{r^2}\frac{d}{dr}\left[r^2\left(\nu_i k_c \frac{dT}{dr} + \Delta H D_i \frac{dc_i}{dr}\right)\right]=0 \tag{2.2.7}$$

which upon integration and use of Eq. (2.2.4) give

$$\nu_j D_i \frac{dc_i}{dr} - \nu_i D_j \frac{dc_j}{dr} = 0, \tag{2.2.8}$$

$$\nu_i k_c \frac{dT}{dr} + \Delta H D_i \frac{dc_i}{dr} = 0. \tag{2.2.9}$$

These relations show that the family of curves

$$\frac{dc_i}{ds} = \frac{\nu_i}{D_i} \qquad i=1,\ldots,N, \tag{2.2.10}$$

$$\frac{dT}{ds} = -\frac{\Delta H}{k_c} \tag{2.2.11}$$

are invariant (or integral) for Eqs. (2.2.2), (2.2.3). We define then as the invariant manifold $\Gamma(u_0)$ the arc in E^+_{N+1} passing through the point $\mathbf{u}_0$ and satisfying Eqs. (2.2.10), (2.2.11).

It can be seen from Eqs. (2.2.10), (2.2.11) that $\Gamma(u_0)$ is a bounded arc under very general assumptions about the transport coefficients $D_1, \dots, D_N, k_c$. Accordingly, one can obtain a priori bounds for the state variables $\mathbf{u}(r)$, depending on $\mathbf{u}_0$ and the transport coefficients but independent of the reaction rates. In the special case of constant $D_1, \dots, D_N, k_c, \Delta H$, Eqs. (2.2.10), (2.2.11) give upon integration

$$c_i - c_{i0} = s\frac{\nu_i}{D_i} \qquad i = 1, \dots, N, \tag{2.2.12}$$

$$T - T_0 = -s\frac{\Delta H}{k_c}. \tag{2.2.13}$$

These two equations define the invariant manifold $\Gamma(u_0)$ as a line segment parametrized by s, $s_1 \leqq s \leqq s_2$. Outside of the segment $\Gamma(u_0)$ one or more of the state variables are negative. The numbers s_1, s_2 and the a priori bounds for $c_1(r), \dots, c_N(r), T(r)$ can be easily obtained from Eqs. (2.2.12), (2.2.13) and the condition $\mathbf{u}(r) \in \Gamma(u_0)$.

Another type of boundary conditions arises when the resistance for heat and mass transfer between the surface of the pellet and the surrounding fluid is not much smaller than the resistance for transport within the pellet. In this case the surface state $\mathbf{u}(r_0)$ is not the same with the state of the surrounding fluid $\mathbf{u}_a$ and the boundary conditions can be formulated by means of *transfer coefficients* $l_1, \dots, l_{N+1}$ as

$$l_i(c_i(r_0) - c_{ia}) = -D_i\left(\frac{dc_i}{dr}\right)_{r_0} \qquad i = 1, \dots, N, \tag{2.2.14}$$

$$l_{N+1}(T(r_0) - T_a) = -k_c\left(\frac{dT}{dr}\right)_{r_0}. \tag{2.2.15}$$

By introducing the boundary conditions (2.2.14), (2.2.15) into Eqs. (2.2.8), (2.2.9) there is obtained

$$\frac{\nu_i}{l_i}(c_i(r_0) - c_{ia}) = \dots = \frac{\nu_N}{l_N}(c_N(r_0) - c_{Na}) = -\frac{\Delta H}{l_{N+1}}(T(r_0) - T_a) \tag{2.2.16}$$

and although the surface point $\mathbf{u}(r_0)$ lies on the straight line of Eq. (2.2.16), its exact location cannot be determined a priori. In this case the invariant manifold of the differential Eqs. (2.2.2), (2.2.3) is two dimensional. It consists of the family of curves (2.2.10), (2.2.11) passing through the points of the straight line (2.2.16).

A priori bounds for the state variables can be derived by choosing i, j such that $\nu_i \nu_j < 0$ and writing Eq. (2.2.16) as

$$\frac{|\nu_i|}{l_i} c_i(r_0) + \frac{|\nu_j|}{l_j} c_j(r_0) = \frac{|\nu_i|}{l_i} c_{ia} + \frac{|\nu_j|}{l_j} c_{ja}. \qquad (2.2.17)$$

Similarly, if $\nu_i \Delta H > 0$, Eq. (2.2.16) gives

$$\frac{|\nu_i|}{l_i} c_i(r_0) + \frac{|\Delta H|}{l_{N+1}} T(r_0) = \frac{|\nu_i|}{l_i} c_{ia} + \frac{|\Delta H|}{l_{N+1}} T_a. \qquad (2.2.18)$$

Eqs. (2.2.17), (2.2.18) provide bounds for the surface variables $\mathbf{u}(r_0)$ which can be used to obtain bounds for $\mathbf{u}(r)$, $0 \leqq r \leqq r_0$.

2.3. A Closed Form Solution

With the exception of linear equations, closed form solutions can be obtained only in the case of a single reaction in a planar region. An example is provided by the irreversible reaction

$$\mathscr{M}_2 - \mathscr{M}_1 = 0 \qquad (2.3.1)$$

with a rate $f(c_1, c_2, T)$ which vanishes at $T=0, c_1=0$ and is positive everywhere else. If the reacting region is bounded by the two parallel planes $x = \pm x_0$ and the transport coefficients are constant, the conservation equations are

$$D_1 \frac{d^2 c_1}{dx^2} = f(c_1, c_2, T), \qquad (2.3.2)$$

$$D_2 \frac{d^2 c_2}{dx^2} = -f(c_1, c_2, T), \qquad (2.3.3)$$

$$k_c \frac{d^2 T}{dx^2} = \Delta H\, f(c_1, c_2, T) \qquad (2.3.4)$$

with boundary conditions:

$$x=0: \quad \frac{dc_1}{dx}=0, \quad \frac{dc_2}{dx}=0, \quad \frac{dT}{dx}=0, \qquad (2.3.5)$$

$$x=x_0: \quad c_1=c_{10}, \quad c_2=c_{20}, \quad T=T_0. \qquad (2.3.6)$$

By eliminating f from the above equations and integrating, there is obtained

$$c_2 = c_{20} - \frac{D_1}{D_2}(c_1 - c_{10}), \qquad (2.3.7)$$

$$T = T_0 + \frac{\Delta H\, D_1}{k_c}(c_1 - c_{10}) \qquad (2.3.8)$$

so that there is only one independent equation

$$D_1 \frac{d^2 c_1}{dx^2} = \tilde{f}(c_1). \tag{2.3.9}$$

As usual, the function $\tilde{f}$ is obtained by introducing Eqs. (2.3.7), (2.3.8) in f. It is convenient to use the dimensionless variables

$$\zeta = \frac{x}{x_0}, \quad \eta = \frac{c_1}{c_{10}}, \quad \phi = \frac{\tilde{f}}{\tilde{f}(c_{10})} \tag{2.3.10}$$

and rewrite Eq. (2.3.9) as

$$\frac{d^2 \eta}{d\zeta^2} = h^2 \phi(\eta), \tag{2.3.11}$$

$$\zeta = 0: \ \frac{d\eta}{d\zeta} = 0, \tag{2.3.12}$$

$$\zeta = 1: \ \eta = 1 \tag{2.3.13}$$

where

$$h = x_0 \left(\frac{\tilde{f}(c_{10})}{D_1 c_{10}} \right)^{\frac{1}{2}}. \tag{2.3.14}$$

AMUNDSON and RAYMOND [2] have investigated the number and stability of solutions of the problem (2.3.11)–(2.3.13). Following their procedure we set $\psi(\eta) = d\eta/d\zeta$ and use η as a new independent variable and ψ as a new dependent variable and obtain from Eq. (2.3.11)

$$\psi(\eta) = h \left\{ 2 \int_{\eta(0)}^{\eta} \phi(\eta')\, d\eta' \right\}^{\frac{1}{2}}. \tag{2.3.15}$$

The above substitution is valid under the condition that $\phi(\eta)$ has a single zero, therefore applies to irreversible as well as reversible reactions.

From Eqs. (2.3.7), (2.3.10), it is seen that η can vary in the interval $[\eta^*, 1]$, where

$$\eta^* = \begin{cases} 0, & \Delta H < 0 \\ \max\left\{0,\ 1 - \dfrac{k_c T_0}{D_1 \Delta H}\right\}, & \Delta H > 0 \end{cases} \tag{2.3.16}$$

is the point of kinetic equilibrium, $\phi(\eta^*) = 0$. Outside the interval $[\eta^*, 1]$ one at least of the variables c_1, T becomes negative.

By integrating Eq. (2.3.15) there is obtained

$$\int_{\eta(0)}^{\eta} \Phi(\eta', \eta(0))\, d\eta' = h\,\zeta \tag{2.3.17}$$

where

$$\Phi(\eta', \eta) = \left\{ 2 \int_{\eta}^{\eta'} \phi(\eta'')\, d\eta'' \right\}^{-\frac{1}{2}}. \tag{2.3.18}$$

The quantity $\eta(0)$ is now determined from the boundary condition $\eta(1) = 1$:

$$J(\eta(0)) = \int_{\eta(0)}^{1} \Phi(\eta, \eta(0))\, d\eta = h. \tag{2.3.19}$$

For fixed h, Eq. (2.3.19) may be satisfied by several values of $\eta(0)$ depending on the nature of the function J. The function J can be studied by isolating its singularity

$$\begin{aligned} J(\eta) = &\int_{\eta}^{1} \{\Phi(\eta', \eta) - [2\phi(\eta)(\eta' - \eta)]^{-\frac{1}{2}}\}\, d\eta' \\ &+ \int_{\eta}^{1} [2\phi(\eta)(\eta' - \eta)]^{-\frac{1}{2}}\, d\eta'. \end{aligned} \tag{2.3.20}$$

Except possibly at $\eta = \eta^*$, $J(\eta)$ has a continuous derivative given by

$$\begin{aligned} \frac{dJ}{d\eta} = &(2\phi(\eta))^{-\frac{1}{2}} \\ &\cdot \left\{ \frac{1}{2} \int_{\eta}^{1} \left[\left(\int_{\eta}^{\eta'} \frac{\phi(\eta'')}{\phi(\eta)}\, d\eta'' \right)^{-\frac{3}{2}} - (\eta' - \eta)^{-\frac{3}{2}} \right] d\eta' - (1 - \eta)^{-\frac{1}{2}} \right\} \end{aligned} \tag{2.3.21}$$

from which it follows easily that

$$\lim_{\eta \to \eta^*} \frac{dJ}{d\eta} < 0. \tag{2.3.22}$$

Fig. 2.3.1a–d shows four types of functions $J(\eta)$. Figs. 2.3.1a, b present the common case in which $\Phi(\eta', \eta)$ has a nonintegrable singularity at $\eta' = \eta = \eta^*$. If J is monotone, there is a single solution for all h. If J is not monotone there are three solutions when h lies in (h_1, h_2) and a single solution when h lies outside that interval. If the function J has more local extrema, a larger and always odd number of solutions obtains in a certain range of values of h. Figs. 2.3.1c, d represent the less common case in which the function $\Phi(\eta', \eta)$ has an integrable singularity at $\eta' = \eta = \eta^*$. Such a singularity may arise, for example, from a rate proportional to $c_1^{\frac{1}{2}}$. Now if $h \leqq J(\eta^*)$ the solutions are obtained as in cases a, b. If, however,

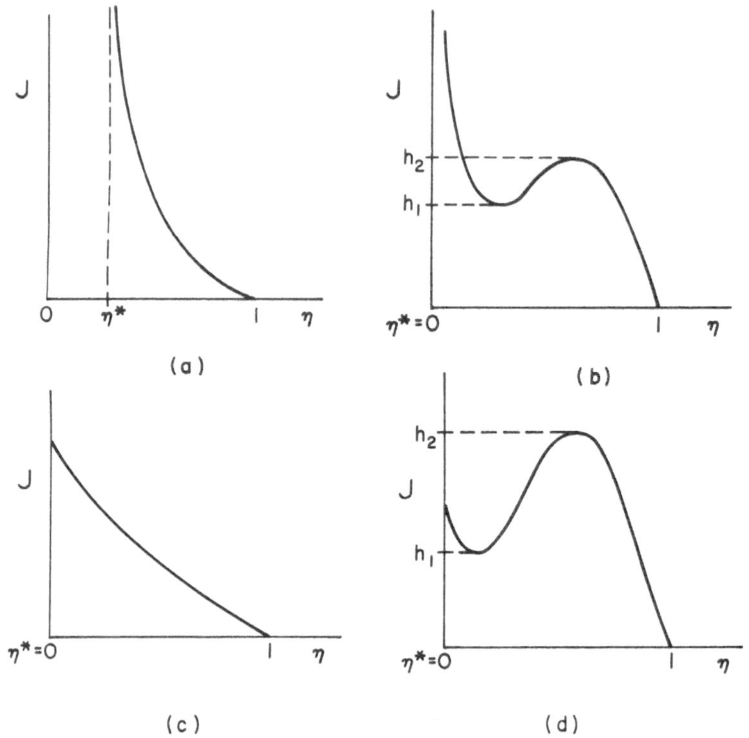

Fig. 2.3.1. The function J of Eq. (2.3.19)

$h > J(\eta^*)$, in addition to the two possible solutions defined by Eqs. (2.3.15), (2.3.17), (2.3.19) there is the solution

$$\eta(\zeta)=\begin{cases} 0, & 0 \leqq \zeta \leqq \zeta_0 \\ \int\limits_{\eta^*}^{\eta} \Phi(\eta,\eta^*)\,d\eta = h(1-\zeta), & \zeta_0 \leqq \zeta \leqq 1 \end{cases} \tag{2.3.23}$$

where

$$\zeta_0 = 1 - \frac{J(\eta^*)}{h}. \tag{2.3.24}$$

In the preceding discussion it was tacitly assumed that h and ϕ can be varied independently. This can be achieved by varying x_0 or by varying the transport coefficients D_1, D_2, k_c by a common factor. On the other hand, variations in the surface conditions c_{10}, c_{20}, T_0 and general variations of D_1, D_2, k_c change both h and ϕ.

In summary, except for situations such as $h=h_1, h=h_2$ in Fig. 2.3.1 b, d, Eq. (2.3.9) has always an odd number of solutions. A necessary and sufficient condition for uniqueness is that $dJ/d\eta < 0, \eta \in (\eta^*, 1]$. This however requires the computation of Eq.(2.3.21) and is rather inconvenient to use. A sufficient condition for uniqueness is $d\tilde{f}/dc_1<0, c_1\in[c_1^*, c_{10}]$. For if $c_1^{(1)}(x), c_1^{(2)}(x)$ are any two solutions of Eq.(2.3.9), the difference $c_1^{(2)}(x)-c_1^{(1)}(x)$ satisfies

$$\frac{d^2}{dx^2}\left(c_1^{(2)}(x)-c_1^{(1)}(x)\right)=\frac{d\tilde{f}(\bar{c}_1(x))}{dc_1}\left(c_1^{(2)}(x)-c_1^{(1)}(x)\right) \qquad (2.3.25)$$

where $\bar{c}_1$ is intermediate between $c_1^{(1)}, c_1^{(2)}$, therefore $c_1^{(2)}-c_1^{(1)}$ cannot have a negative minimum or a positive maximum. The steady state is also unique for sufficiently small and sufficiently large values of h, for example when the catalyst pellet is very small or very large. The important dimensionless parameter h is known in the chemical and engineering literature as the Thiele Modulus. Note the similar role played by the quantity $h^2=x_0^2\tilde{f}(c_{10})/D_1c_{10}$ and the parameter $V\tilde{f}_i(\mathbf{u}_f)/w\,c_i$ encountered in Section 1.8.

Effectiveness Factor and Asymptotic Behavior

In engineering work it is desirable to express the total reaction rate in a catalyst pellet in terms of the surface conditions alone. For this reason, the effect of the concentration and temperature profiles within the particle are incorporated in a single quantity, the *effectiveness factor*. For the case of a single reaction, the effectiveness factor is defined as:

$$\mathsf{E}=\frac{\int_V f(\mathbf{u}(r))\,dv(r)}{|V|\,f(\mathbf{u}_0)} \qquad (2.3.26)$$

where $\mathbf{u}=(c_1,\dots,c_N,T)$, V is the reacting region, $|V|$ its volume, and $\mathbf{u}_0$ are the surface conditions. For plane and spherical regions, the effectiveness factor is simply

$$\mathsf{E}^{(1)}=\frac{\int_0^{x_0} f(\mathbf{u}(x))\,dx}{x_0\,f(\mathbf{u}_0)}, \qquad (2.3.27)$$

$$\mathsf{E}^{(3)}=\frac{3\int_0^{r_0} r^2 f(\mathbf{u}(r))\,dr}{r_0^3\,f(\mathbf{u}_0)}. \qquad (2.3.28)$$

In the simple example studied in this section the effectiveness factor is

$$\mathsf{E}^{(1)}=\frac{\int_0^{x_0} \tilde{f}(c_1(x))\,dx}{x_0\,\tilde{f}(c_{10})}. \qquad (2.3.29)$$

This can be evaluated by means of Eqs.(2.3.9) and (2.3.14) as

$$\mathrm{E}^{(1)} = \frac{D_1\left(\dfrac{dc_1}{dx}\right)_{x=x_0}}{x_0 \tilde{f}(c_{10})} = \frac{1}{h}\left[2\int_{\eta(0)}^{1} \phi(\eta')\,d\eta'\right]^{\frac{1}{2}}. \qquad (2.3.30)$$

As $h \to 0$, $\tilde{f}(c_1) \to \tilde{f}(c_{10})$ and $\mathrm{E}^{(1)} \to 1$. As $h \to \infty$, $\eta(0) \to \eta^*$ and

$$\mathrm{E}^{(1)} \to \frac{1}{h}\left[2\int_{\eta^*}^{1} \phi(\eta')\,d\eta'\right]^{\frac{1}{2}}. \qquad (2.3.31)$$

This formula is useful in practice because for many reactions $\mathrm{E}^{(1)}$ is very close to its asymptotic value when $h > 1$.

Let us now discuss the physical significance of the Thiele Modulus h. The quantity h^2 is the ratio of the terms x_0^2/D_1 and $c_{10}/\tilde{f}(c_{10})$ which represent "resistances" for diffusion and chemical reaction. Small h indicates large resistance for chemical reaction relative to the resistance for transport. Therefore, the state $\mathbf{u}(x)$ is close to $\mathbf{u}_0$ in the whole region $-x_0 \leqq x \leqq x_0$. Moreover, the reaction rate $\tilde{f}(c_1)$ is close to $\tilde{f}(c_{10})$ so that $|\mathrm{E}^{(1)} - 1| \ll 1$. Large h indicates large resistance for transport relative to the resistance for reaction. In this case the interior of the particle is at a state near kinetic equilibrium due to the fact that diffusion is not large enough to equalize the concentrations. Accordingly, the effectiveness factor is small. This is the *transport limited regime.*

To obtain a more precise understanding of the transport limited regime we write Eq.(2.3.9) as

$$\begin{aligned} D_1\left(\frac{dc_1}{dx}\right)_{x=x_0} &= \int_0^{x_0} \tilde{f}(c_1)\,dx \\ &= \tilde{f}(c_{10})\left[\frac{c_{10}\tilde{f}(c_{10})}{D_1}\right]^{\frac{1}{2}}\left[2\int_{\eta(0)}^{1} \phi(\eta')\,d\eta'\right]^{\frac{1}{2}}. \end{aligned} \qquad (2.3.32)$$

This shows that the total reaction rate in the region $0 \leqq x \leqq x_0$ (per unit area perpendicular to the x direction) is bounded independently of the thickness x_0 of the region, for the integral in the brackets cannot exceed its maximum value obtained for $\eta(0) = \eta^*$. Note also that c_1 is monotone decreasing from the boundary inwards and the quantity

$$\left[2\int_{\eta(0)}^{1} \phi(\eta')\,d\eta'\right]^{\frac{1}{2}}$$

is of the order of unity. Hence, Eq.(2.3.32) also shows that the reaction essentially takes place only in a surface layer of thickness $[D_1 c_{10}/\tilde{f}(c_{10})]^{\frac{1}{2}} = x_0/h$. The remaining region in the interior is very near the kinetic equilibrium state $c_1^* = \eta^* c_{10}$.

PETERSEN [34] has extended the analysis of the transport limited regime to spherical regions. For a spherical region, Eq. (2.3.9) has the form

$$\frac{1}{r^2}\frac{d}{dr}\left(r^2\frac{dc_1}{dr}\right)=\tilde{f}(c_1), \tag{2.3.33}$$

$$r=0:\quad \frac{dc_1}{dr}=0, \tag{2.3.34}$$

$$r=r_0:\quad c_1=c_{10}. \tag{2.3.35}$$

and for large h, the reaction goes to completion in a thin surface layer. Following PETERSEN we use the dimensionless variables

$$\rho=1-\frac{r}{r_0},\quad \eta=\frac{c_1}{c_{10}},\quad \phi(\eta)=\frac{\tilde{f}(c_1)}{\tilde{f}(c_{10})}, \tag{2.3.36}$$

$$h=r_0\left(\frac{\tilde{f}(c_{10})}{D_1\,c_{10}}\right)^{\frac{1}{2}} \tag{2.3.37}$$

to write Eqs. (2.3.33) – (2.3.35) as

$$\frac{d^2\eta}{d\rho^2}-\frac{2}{1-\rho}\frac{d\eta}{d\rho}=h^2\,\phi(\eta), \tag{2.3.38}$$

$$\rho=1:\quad \frac{d\eta}{d\rho}=0, \tag{2.3.39}$$

$$\rho=0:\quad \eta=1. \tag{2.3.40}$$

In terms of ρ the thickness of the reacting surface layer is of the order of $1/h$. Thus we can make the order of magnitude estimates

$$\frac{d^2\eta}{d\rho^2}\sim O(h^2),\quad \frac{2}{1-\rho}\frac{d\eta}{d\rho}\sim O(h),\quad h^2\,\phi(\eta)\sim O(h^2) \tag{2.3.41}$$

and neglect the second term in Eq. (2.3.38) to obtain

$$\frac{d^2\eta}{d\rho^2}=h^2\,\phi(\eta). \tag{2.3.42}$$

Also, since the interior of the region is very near equilibrium we can replace the boundary condition (2.3.39) by

$$\rho=1:\ \eta=\eta^*. \tag{2.3.43}$$

Eqs. (2.3.42), (2.3.43), (2.3.40) can be integrated to

$$\frac{d\eta}{d\rho}=-h\left[2\int_{\eta^*}^{\eta}\phi(\eta')\,d\eta'\right]^{\frac{1}{2}}. \tag{2.3.44}$$

From this expression and Eqs.(2.3.26), (2.3.42) we can calculate total reaction rate and effectiveness factor as

$$4\pi \int_0^{r_0} r^2 \tilde{f}(c_1(r))\, dr = 4\pi\, r_0^2\, \tilde{f}(c_{10}) \frac{r_0}{h} \left[2 \int_{\eta^*}^{1} \phi(\eta')\, d\eta' \right]^{\frac{1}{2}}, \qquad (2.3.45)$$

$$\mathrm{E}^{(3)} = \frac{3}{h} \left[2 \int_{\eta^*}^{1} \phi(\eta')\, d\eta' \right]^{\frac{1}{2}}. \qquad (2.3.46)$$

2.4. General Reaction Systems. Invariant Manifolds and A Priori Bounds

From this point and on it will be assumed that the transport coefficients $D_1, \ldots, D_N, k_c$ are constant so that Eqs.(2.1.34), (2.1.35) simplify to

$$D_i \frac{1}{r^2} \frac{d}{dr}\left(r^2 \frac{dc_i}{dr}\right) = -\sum_{j=1}^{R} v_{ij} f_j \qquad i = 1, \ldots, N, \qquad (2.4.1)$$

$$k_c \frac{1}{r^2} \frac{d}{dr}\left(r^2 \frac{dT}{dr}\right) = \sum_{j=1}^{R} \Delta H_j f_j. \qquad (2.4.2)$$

The boundary conditions to be considered initially are

$$r=0: \quad \frac{dc_i}{dr} = 0, \qquad \frac{dT}{dr} = 0, \qquad (2.4.3)$$

$$r=r_0: \quad c_i = c_{i0}, \qquad T = T_0. \qquad (2.4.4)$$

The procedure of obtaining the invariant manifolds is similar to that of Section 1.2. By multiplying Eq.(2.4.1) by γ_{il}, summing over i, integrating twice, and using Eqs.(2.4.3), (2.4.4) there is obtained

$$\sum_{i=1}^{N} \gamma_{il} D_i (c_i - c_{i0}) = 0 \qquad l = 1, \ldots, N-R \qquad (2.4.5)$$

where the significance of the coefficients γ_{il} is discussed in Section 1.2. Recalling the definition

$$\Delta H_j = \sum_{i=1}^{N} v_{ij} H_i$$

and the assumption of constant ΔH_j, we can assume with no further loss of generality that the enthalpies H_i are also constants. Multiplying then Eq.(2.4.1) by H_i, summing over i, adding to Eq.(2.4.2), and integrating twice there is obtained

$$k_c (T - T_0) + \sum_{i=1}^{N} D_i H_i (c_i - c_{i0}) = 0. \qquad (2.4.6)$$

The set of points in E_{N+1}^{+} that satisfy Eqs.(2.4.5), (2.4.6) will be denoted by $\Gamma(u_0)$ and called the invariant manifold containing the point $\mathbf{u}_0$.

The region E_{N+1}^{+} is decomposed to an infinite number of mutually exclusive invariant manifolds. Two manifolds having a common point coincide.

It follows from its definition that the manifold $\Gamma(u_0)$ is closed and convex. It is also bounded as can be seen by setting $\gamma_{il}=\beta_{il}>0$, $l=1, \ldots, R_\beta$ in Eq.(2.4.5) to obtain

$$\sum_{i=1}^{N} \beta_{il} D_i(c_i - c_{i0}) = 0 \qquad l=1, \ldots, R_\beta \tag{2.4.7}$$

where the significance of the constants γ_{il}, β_{il} has been given in Section 1.2.

Due to the linearity of Eqs.(2.4.6), (2.4.7), the maximum and minimum of each state variable in $\Gamma(u_0)$ is attained at the intersection of $\Gamma(u_0)$ with the surface ∂E_{N+1}^{+}, at points where one or more of the state variables vanish. These maxima and minima, which depend on the boundary condition $\mathbf{u}_0$ and the transport coefficients but are independent of the reaction rates and the size r_0 of the region, are *a priori bounds* for the solutions of the boundary value problem (2.4.1)–(2.4.4). The independence of the a priori bounds of the reaction rates and the size of the region has very important mathematical implications and will be used extensively in the following sections.

As with uniform time dependent systems, each invariant manifold can be described by R coordinates $\xi_1, \ldots, \xi_R$, the extents of reactions, defined by:

$$D_i(c_i - c_{i0}) = \sum_{j=1}^{R} \nu_{ij} \xi_j \qquad i=1, \ldots, N, \tag{2.4.8}$$

$$k_c(T - T_0) = -\sum_{j=1}^{R} \Delta H_j \xi_j . \tag{2.4.9}$$

Eqs.(2.4.8), (2.4.9) and their inverse will be occasionally denoted as

$$\mathbf{u} = \mathbf{u}(\boldsymbol{\xi}; \mathbf{u}_0), \tag{2.4.10}$$

$$\boldsymbol{\xi} = \boldsymbol{\xi}(\mathbf{u}; \mathbf{u}_0). \tag{2.4.11}$$

They establish a one-one correspondence between the region $\Gamma(u_0)$ and its image $\tilde{\Gamma}(u_0)$ in the $\boldsymbol{\xi}$ space:

$$\tilde{\Gamma}(u_0) = \{\boldsymbol{\xi} : \boldsymbol{\xi}(\mathbf{u}; \mathbf{u}_0), \mathbf{u} \in \Gamma(\mathbf{u}_0)\} . \tag{2.4.12}$$

The region $\tilde{\Gamma}(u_0)$ is bounded, closed, and convex. Its boundary $\partial\tilde{\Gamma}(u_0)$ consists of points where one or more of the state variables $c_1, \ldots, c_N$, T vanishes.

As usual we shall set

$$\tilde{f}_j(\boldsymbol{\xi}) = f_j(\mathbf{u}(\boldsymbol{\xi}; \mathbf{u}_0)) \qquad j=1, \ldots, R \tag{2.4.13}$$

and obtain Eqs. (2.4.1)–(2.4.4) in the reduced form

$$\frac{1}{r^2}\frac{d}{dr}\left(r^2\frac{d\xi_j}{dr}\right)=-\tilde{f}_j(\xi_1,\dots,\xi_R) \qquad j=1,\dots,R, \tag{2.4.14}$$

$$r=0:\ \frac{d\xi_j}{dr}=0, \tag{2.4.15}$$

$$r=r_0:\quad \xi_j=0. \tag{2.4.16}$$

Boundary Conditions of the Third Kind

The situation is more complicated with boundary conditions of the third kind:

$$r=0:\ \frac{dc_i}{dr}=0,\quad \frac{dT}{dr}=0, \tag{2.4.17}$$

$$r=r_0:\ l_i(c_i-c_{ia})=-D_i\left(\frac{dc_i}{dr}\right)_{r=r_0},\qquad l_{N+1}(T-T_a)=-k_c\left(\frac{dT}{dr}\right)_{r=r_0}. \tag{2.4.18}$$

Again, by eliminating the rates f_j from Eqs. (2.4.1)–(2.4.2) (as with the derivation of Eqs. (2.4.5), (2.4.6)), integrating and using Eqs. (2.4.17), (2.4.18) there is obtained

$$\sum_{i=1}^{N} l_i\gamma_{ik}(c_i(r_0)-c_{ia})=0 \qquad k=1,\dots,N-R, \tag{2.4.19}$$

$$l_{N+1}(T(r_0)-T_a)+\sum_{i=1}^{N} l_i H_i(c_i(r_0)-c_{ia})=0. \tag{2.4.20}$$

Eqs. (2.4.19), (2.4.20) define an R-dimensional manifold whose restriction in E^+_{N+1} will be denoted by $\Gamma_1(u_a)$. The manifold $\Gamma_1(u_a)$ would be identical to $\Gamma(u_a)$ if $l_1,\dots,l_{N+1}$ were replaced by $D_1,\dots,D_N,k_c$. Since the manifold $\Gamma_1(u_a)$ is bounded, a priori bounds can be obtained for the surface conditions $\mathbf{u}(r_0)\in\Gamma_1(u_a)$ and hence for $\mathbf{u}(r)$ as well, $0\leqq r\leqq r_0$. Any steady state $\mathbf{u}(r)$, i.e., any solution of Eqs. (2.4.1)–(2.4.4), lies entirely in the manifold

$$\Gamma_2(u_a)=\bigcup_{\mathbf{v}\in\Gamma_1(u_a)}\Gamma(v). \tag{2.4.21}$$

The dimensionality of $\Gamma_2(u_a)$ can be determined as follows. A set of the type $\Gamma(v)\cap\Gamma_1(u_a)$, $\mathbf{v}\in\Gamma(u_a)$, consists of all points $\mathbf{v}$ which satisfy the equations

$$\sum_{i=1}^{N}\gamma_{il}D_i(v_i-c_{ia})=0 \qquad l=1,\dots,N-R, \tag{2.4.22}$$

$$k_c(v_{N+1}-T_a)+\sum_{i=1}^{N}D_iH_i(v_i-c_{ia})=0, \tag{2.4.23}$$

$$\sum_{i=1}^{N}\gamma_{ik}l_i(v_i-c_{ia})=0 \qquad k=1,\dots,N-R, \tag{2.4.24}$$

$$l_{N+1}(v_{N+1}-T_a)+\sum_{i=1}^{N}l_iH_i(v_i-c_{ia})=0 \tag{2.4.25}$$

and therefore has dimension equal to the nullity N_1 of the partitioned matrix $[\gamma/\gamma_1]$, where

$$\gamma=\begin{bmatrix} \gamma_{11} & \gamma_{21} & \cdots & \gamma_{N1} & 0 \\ \gamma_{12} & \gamma_{22} & \cdots & \gamma_{N2} & 0 \\ \cdots & \cdots & \cdots & \cdots & \cdots \\ \gamma_{1,N-R} & \gamma_{2,N-R} & \cdots & \gamma_{N,N-R} & 0 \\ H_1 & H_2 & \cdots & H_N & 1 \end{bmatrix}, \tag{2.4.26}$$

$$\gamma_1=\begin{bmatrix} \frac{l_1}{D_1}\gamma_{11} & \frac{l_2}{D_2}\gamma_{21} & \cdots & \frac{l_N}{D_N}\gamma_{N1} & 0 \\ \frac{l_1}{D_1}\gamma_{12} & \frac{l_2}{D_2}\gamma_{22} & \cdots & \frac{l_N}{D_N}\gamma_{N2} & 0 \\ \cdots & \cdots & \cdots & \cdots & \cdots \\ \frac{l_1}{D_1}\gamma_{1,N-R} & \frac{l_2}{D_2}\gamma_{2,N-R} & \cdots & \frac{l_N}{D_N}\gamma_{N,N-R} & 0 \\ \frac{l_1}{D_1}H_1 & \frac{l_2}{D_2}H_2 & \cdots & \frac{l_N}{D_N}H_N & \frac{l_{N+1}}{k_c} \end{bmatrix}. \tag{2.4.27}$$

In the special case $l_1/D_1=\cdots=l_N/D_N=l_{N+1}/k_c$, the nullity N_1 is equal to R due to the fact that the rows of the matrix γ_1 are multiples of the rows of the matrix γ which has nullity R. In all other cases $N_1=\max\{0, 2R-N-1\}$. Now, since the set $\Gamma(v)\cap\Gamma_1(u_a)$, $\mathbf{v}\in\Gamma(u_a)$, is N_1-dimensional, it can be completely characterized by $R-N_1$ parameters, in addition to the known parameters $\mathbf{u}_a$. Each point $\mathbf{u}\in\Gamma_2(u_a)$ can then be completely characterized by first specifying a set $\Gamma(v)\cap\Gamma_1(u_a)$, $\mathbf{v}\in\Gamma(u_a)$ and then specifying $\mathbf{u}$ by R variables within $\Gamma(v)$. Thus, $\Gamma_2(u_a)$ is $2R-N_1$ dimensional. If $N_1=2R-N-1$, $\Gamma_2(u_a)$ is $N+1$ dimensional and no reduction in the number of differential equations is possible. Of course, a priori bounds can be obtained in all cases.

Example 2.4.1. Consider the two isothermal reactions

$$-\mathscr{M}_1+\mathscr{M}_2=0, \tag{2.4.28}$$

$$-\mathscr{M}_1+\mathscr{M}_3=0. \tag{2.4.29}$$

The invariant manifold corresponding to a point (c_{10}, c_{20}, c_{30}) is shown in Fig. 2.4.1; it is the restriction in E_3^+ of the plane

$$\sum_{i=1}^{3} D_i(c_i-c_{i0})=0. \tag{2.4.30}$$

The a priori bounds are given by

$$0\leqq c_i\leqq\frac{D_1 c_{10}+D_2 c_{20}+D_3 c_{30}}{D_i} \qquad i=1,2,3. \tag{2.4.31}$$

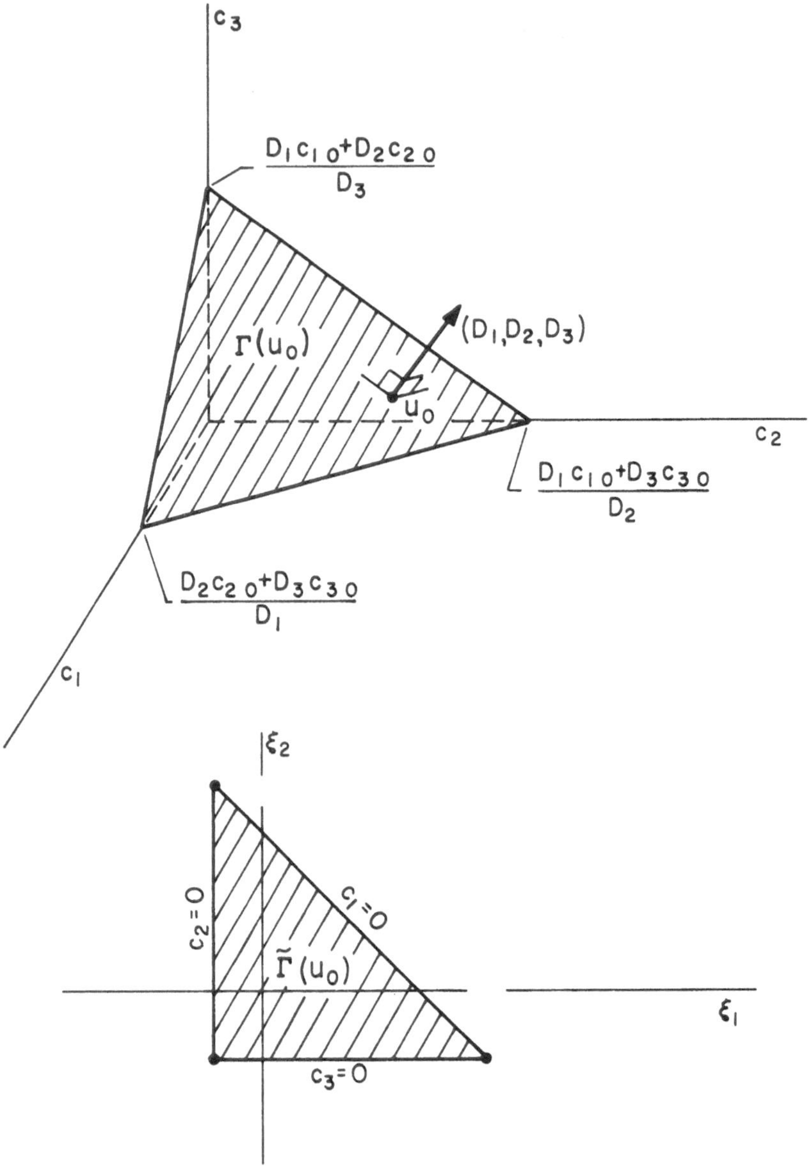

Fig. 2.4.1. An invariant manifold in Example 2.4.1

Fig. 2.4.1 also shows $\tilde{\Gamma}(u_0)$, the image of $\Gamma(u_0)$ in the ξ space, obtained by using Eqs. (2.4.8), (2.4.9).

Example 2.4.2. The second example treats the reaction system

$$-C_2H_4-\tfrac{1}{2}O_2+C_2H_4O=0, \tag{2.4.32}$$

$$-C_2H_4O-\tfrac{5}{2}O_2+2CO_2+2H_2O=0. \tag{2.4.33}$$

There are six state variables, the concentrations c_1, c_2, c_3, c_4, c_5 of the species $C_2H_4, C_2H_4O, O_2, CO_2, H_2O$ and the temperature T. The steady state distributions of the state variables in a spherical catalyst pellet of radius r_0 are described by the boundary value problem

$$D_1\frac{d}{dr}\left(r^2\frac{dc_1}{dr}\right)=f_1, \tag{2.4.34}$$

$$D_2\frac{d}{dr}\left(r^2\frac{dc_2}{dr}\right)=-f_1+f_2, \tag{2.4.35}$$

$$D_3\frac{d}{dr}\left(r^2\frac{dc_3}{dr}\right)=\frac{1}{2}f_1+\frac{5}{2}f_2, \tag{2.4.36}$$

$$D_4\frac{d}{dr}\left(r^2\frac{dc_4}{dr}\right)=-2f_2, \tag{2.4.37}$$

$$D_5\frac{d}{dr}\left(r^2\frac{dc_5}{dr}\right)=-2f_2, \tag{2.4.38}$$

$$k_c\frac{d}{dr}\left(r^2\frac{dT}{dr}\right)=\Delta H_1 f_1+\Delta H_2 f_2, \tag{2.4.39}$$

$$r=0:\ \frac{dc_i}{dr}=0 \quad (i=1,\dots,5), \quad \frac{dT}{dr}=0, \tag{2.4.40}$$

$$r=r_0:\ c_i=c_{i\,0} \quad (i=1,\dots,5), \quad T=T_0. \tag{2.4.41}$$

The two dimensional invariant manifolds ($R=2$) are defined by four equations. These can be obtained from Eqs.(2.4.34)–(2.4.39) by eliminating the rates f_1, f_2 and integrating. Alternatively, they can be obtained by writing the conservation equations for the atomic species O, C, H and the energy:

$$D_2(c_2-c_{2\,0})+2D_3(c_3-c_{3\,0})+2D_4(c_4-c_{4\,0})+D_5(c_5-c_{5\,0})=0, \tag{2.4.42}$$

$$2D_1(c_1-c_{1\,0})+2D_2(c_2-c_{2\,0})+D_4(c_4-c_{4\,0})=0, \tag{2.4.43}$$

$$2D_1(c_1-c_{1\,0})+2D_2(c_2-c_{2\,0})+D_5(c_5-c_{5\,0})=0, \tag{2.4.44}$$

$$\sum_{i=1}^{5}H_i D_i(c_i-c_{i\,0})+k_c(T-T_0)=0. \tag{2.4.45}$$

The extents ξ_1, ξ_2 are defined by

$$D_1(c_1 - c_{1\,0}) = -\xi_1, \tag{2.4.46}$$

$$D_2(c_2 - c_{2\,0}) = \xi_1 - \xi_2, \tag{2.4.47}$$

$$D_3(c_3 - c_{3\,0}) = -\tfrac{1}{2}\xi_1 - \tfrac{5}{2}\xi_2, \tag{2.4.48}$$

$$D_4(c_4 - c_{4\,0}) = 2\xi_2, \tag{2.4.49}$$

$$D_5(c_5 - c_{5\,0}) = 2\xi_2, \tag{2.4.50}$$

$$k_c(T - T_0) = -\Delta H_1\,\xi_1 - \Delta H_2\,\xi_2. \tag{2.4.51}$$

The invariant manifold $\tilde{\Gamma}(u_0)$ is the convex two dimensional region defined by the inequalities

$$-\xi_1 \geqq -D_1\,c_{1\,0}, \tag{2.4.52}$$

$$\xi_1 - \xi_2 \geqq -D_2\,c_{2\,0}, \tag{2.4.53}$$

$$-\tfrac{1}{2}\xi_1 - \tfrac{5}{2}\xi_2 \geqq -D_3\,c_{3\,0}, \tag{2.4.54}$$

$$2\xi_2 \geqq -D_4\,c_{4\,0}, \tag{2.4.55}$$

$$2\xi_2 \geqq -D_5\,c_{5\,0}, \tag{2.4.56}$$

$$-\Delta H_1\,\xi_1 - \Delta H_2\,\xi_2 \geqq -k_c\,T_0. \tag{2.4.57}$$

At each point of the closed polygonal line $\partial\tilde{\Gamma}(u_0)$ one or more of the state variables $c_1, \ldots, c_5, T$ vanishes. The exact shape of this boundary depends on the relative magnitude of the quantities at the right of Eqs. (2.4.52)–(2.4.57). Fig. 2.4.2 shows an example of a manifold $\tilde{\Gamma}(u_0)$. The a priori bounds can be easily obtained from the manifold $\tilde{\Gamma}(u_0)$ and Eqs. (2.4.46)–(2.4.51). For example, the maximum of c_1, attained at the point A, is $c_{1\,0} + (D_2/D_1)\,c_{2\,0} + (D_5/2D_1)\,c_{5\,0}$, and the minimum of c_1, attained at the point B, is zero. At higher dimensions the determination of a priori bounds in essentially a simple problem in linear programming.

The manifolds considered so far are entirely independent of the reaction rates. Limited information about the reaction rates can be utilized to define an *accessible* part of the manifold and thereby obtain sharper a priori bounds. If both reactions in Example 2.4.2 are irreversible, ξ_1 and ξ_2 cannot take negative values and the accessible part of the manifold $\tilde{\Gamma}(u_0)$ lies in the positive quadrant of Fig. 2.4.2. The utilization of limited information about the reaction rates in defining an accessible part within

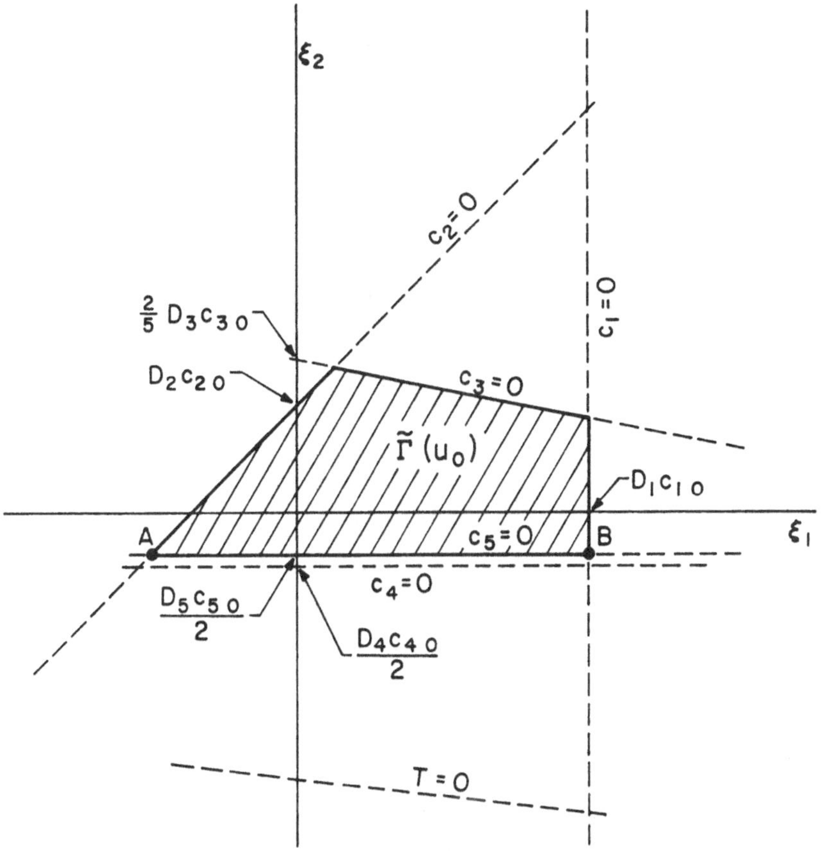

Fig. 2.4.2. An invariant manifold in Example 2.4.2

a manifold is a rather interesting subject which has not yet been investigated systematically.

2.5. Existence of Solutions

In this section we shall prove the existence of steady states, i.e., the existence of solutions for the boundary value problem (2.4.1)–(2.4.4), or its equivalent (2.4.14)–(2.4.16). The geometrical structure of this problem is very similar to that of Eqs. (1.8.27), (1.8.28), and use can be made of precisely the same fixed point method used in Section 1.8.

Before defining the basic function space and operator, we would like to exclude the possibility of spurious solutions with negative components which may interfere with the existence proof. The following artificial example shows that spurious solutions may indeed result from a nondiscriminating extension of the rates f_j in the space of negative concentrations.

$$-\mathcal{M}_1+\mathcal{M}_2=0, \tag{2.5.1}$$

$$f(c_1)=|c_1|, \quad -\infty<c_1<\infty, \tag{2.5.2}$$

$$\frac{d^2 c_1}{dx^2}=|c_1|, \tag{2.5.3}$$

$$x=0:\ \frac{dc_1}{dx}=0; \quad x=\pi:\ c_1=\sinh\frac{\pi}{2}. \tag{2.5.4}$$

The above boundary value problem has the nonnegative solution

$$c_1(x)=\frac{\sinh\frac{\pi}{2}}{\cosh\pi}\cosh x, \quad 0\leqq x\leqq\pi \tag{2.5.5}$$

and the spurious solution

$$c_1(x)=\begin{cases}-\cos x, & 0\leqq x\leqq\frac{\pi}{2}\\ \sinh\left(x-\frac{\pi}{2}\right), & \frac{\pi}{2}\leqq x\leqq\pi\end{cases} \tag{2.5.6}$$

which arises from the physically meaningless definition of f in $(-\infty,0)$. Of the two solutions (2.5.5), (2.5.6), only the first is the asymptotic limit, as $t\to\infty$ of the solution of the time dependent problem

$$\frac{\partial c_1}{\partial t}=\frac{\partial^2 c_1}{\partial x^2}+f(c_1) \tag{2.5.7}$$

with boundary conditions (2.5.4) and nonnegative initial conditions.

The spurious solutions are avoided by the special definition

$$\sum_{j=1}^{R}\nu_{ij}f_j(c_1,\dots,c^-,\dots,T)=\sum_{j=1}^{R}\nu_{ij}f_j(c_1,\dots,0,\dots,T)+D_i(c_{i0}-c_i)d(\mathbf{u}), \tag{2.5.8}$$

$$\sum_{j=1}^{R}\Delta H_j f_j(c_1,\dots,c^-,\dots,T)=\sum_{j=1}^{R}\Delta H_j f_j(c_1,\dots,0,\dots,T)+k_c(T_0-T)d(\mathbf{u}) \tag{2.5.9}$$

with the same notation as in Sections 1.4, 1.8. This definition is restricted to points $\mathbf{u}$ satisfying Eqs. (2.4.5), (2.4.6), but this suffices for our purposes. Now if $c_i(r)$ takes on negative values, it attains its negative minimum at an interior point where $dc_i/dr=0$, $d^2 c_i/dr^2\geqq 0$. This, however, contradicts Eq. (2.4.1), which at the point of the minimum gives

$$\frac{d^2 c_i}{dr^2}=-\left\{\sum_{j=1}^{R}\nu_{ij}f_j(c_1,\dots,0,\dots,T)+D_i(c_{i0}-c_i)\,d(\mathbf{u})\right\}. \tag{2.5.10}$$

A similar contradiction results from the assumption $T(r)<0$.

Proceeding to the main topic of the present section let $\mathscr{B}_c$ denote the Banach space of all real continuous vector valued functions

$$\xi(r)=(\xi_1(r),\dots,\xi_R(r)) \tag{2.5.11}$$

with norm

$$\|\xi\|=\sum_{j=1}^{R}\max_{0\leqq r\leqq r_0}|\xi_j(r)|. \tag{2.5.12}$$

Using the GREEN'S function

$$G(r,r')=\begin{cases}\dfrac{r_0-r'}{r'\,r_0}, & 0\leqq r\leqq r'\\[2ex] \dfrac{r_0-r}{r\,r_0}, & r'\leqq r\leqq r_0\end{cases} \tag{2.5.13}$$

we can rewrite Eqs. (2.4.14) – (2.4.16) as

$$\xi(r)=\int_0^{r_0} r'^2\,G(r,r')\,\tilde{f}(\xi(r'))\,dr' \tag{2.5.14}$$

or more succinctly as

$$\xi=\mathbf{H}\,\xi \tag{2.5.15}$$

where $\mathbf{H}$ is the nonlinear operator

$$\mathbf{H}\,\xi=\int_0^{r_0} r'^2\,G(r,r')\,\tilde{f}(\xi(r'))\,dr'. \tag{2.5.16}$$

It can be easily proved by using ASCOLI'S Theorem that $\mathbf{H}$ is completely continuous in $\mathscr{B}_c$.

If ξ is a solution of Eq. (2.5.15), the R-dimensional vector $\xi(r)$ lies in $\tilde{\Gamma}(u_0)$ for all r, therefore ξ as an element of $\mathscr{B}_c$ satisfies

$$\xi\in V(u_0) \tag{2.5.17}$$

where

$$V(u_0)=\{\xi:\ \xi(r)\in\tilde{\Gamma}(u_0),\ 0\leqq r\leqq r_0\} \tag{2.5.18}$$

is a closed bounded convex region in $\mathscr{B}_c$. If

$$S_1=\{\xi:\ \|\xi\|=b\} \tag{2.5.19}$$

is a sphere in $\mathscr{B}_c$ surrounding $V(u_0)$ then no solution of Eq. (2.5.15) can lie on S_1. We are ready now to proceed with the existence proof by showing, as in previous occasions, that the vector fields $\mathbf{I}$ and $\mathscr{H}=\mathbf{I}-\mathbf{H}$ are homotopic on S_1. The "bridging" vector field $\mathscr{H}(s)=s(\mathbf{I}-\mathbf{H})+(1-s)\,\mathbf{I}$, $0\leqq s\leqq 1$, has no null point on S_1, for any such null point would imply that the equation

$$\xi=s\,\mathbf{H}\,\xi \tag{2.5.20}$$

or its equivalent

$$\frac{1}{r^2}\frac{d}{dr}\left(r^2\frac{d\xi}{dr}\right)=-s\,\tilde{f}(\xi), \tag{2.5.21}$$

$$r=0:\ \frac{d\xi}{dr}=0, \tag{2.5.22}$$

$$r=r_0:\ \xi=0 \tag{2.5.23}$$

has a solution $\xi\in S_1$. This contradicts the fact that any solution $\xi(r)$ of Eqs. (2.5.21)–(2.5.23) must lie in $V(u_0)$ for all $s\geqq 0$, since $\tilde{\Gamma}(u_0)$ and $V(u_0)$ do not depend on the magnitude of the reaction rates. The vector fields **I**, **I**$-$**H** as homotopic have the same rotation, $+1$, therefore the operator **H** has at least one fixed point.

Existence of solutions can also be shown for the case of boundary conditions (2.4.17), (2.4.18). This time the reduction of Eqs. (2.4.1), (2.4.2) to R equations is not generally possible and the boundary value problem is written in integral form as

$$c_i(r)=c_{ia}+\int_0^{r_0} r'^2\, G_i(r,r')\sum_{j=1}^{R}\nu_{ij} f_j(\mathbf{u}(r'))\,dr' \qquad i=1,\dots,N, \tag{2.5.24}$$

$$T(r)=T_a-\int_0^{r_0} r'^2\, G_T(r,r')\sum_{j=1}^{R}\Delta H_j f_j(\mathbf{u}(r'))\,dr' \tag{2.5.25}$$

where

$$G_i(r,r')=\begin{cases}\dfrac{1}{r'}+\left(\dfrac{D_i}{l_i r_0^2}-\dfrac{1}{r_0}\right), & 0\leqq r\leqq r'\\[2ex] \dfrac{1}{r}+\left(\dfrac{D_i}{l_i r_0^2}-\dfrac{1}{r_0}\right), & r'\leqq r\leqq r_0,\end{cases} \tag{2.5.26}$$

$$G_T(r,r')=\begin{cases}\dfrac{1}{r'}+\left(\dfrac{k_c}{l_{N+1} r_0^2}-\dfrac{1}{r_0}\right), & 0\leqq r\leqq r'\\[2ex] \dfrac{1}{r}+\left(\dfrac{k_c}{l_{N+1} r_0^2}-\dfrac{1}{r_0}\right), & r'\leqq r\leqq r_0.\end{cases} \tag{2.5.27}$$

Due to the fact that $\mathbf{u}(r)$ does not lie in an a priori determined R-dimensional manifold, the definition (2.5.8), (2.5.9) should be modified. One suitable modification preventing the existence of spurious solutions of the boundary value problem (2.4.1), (2.4.2), (2.4.17), (2.4.18) is

$$f_j(c_1,\dots,c^-,\dots,T)=f_j(c_1,\dots,0,\dots,T) \qquad j=1,\dots,R. \tag{2.5.28}$$

Incidentally, Eq. (2.5.28) can also be used in place of Eqs. (2.5.8), (2.5.9).

The existence proof for Eqs. (2.5.24), (2.5.25) is very similar to that given for Eqs. (2.5.16), except that the equation $\mathscr{H}(s)\,\mathbf{u}=0$ is equivalent to Eqs. (2.4.1), (2.4.2), (2.4.17), (2.4.18) with f_j replaced by $s\,f_j$ and $\mathbf{u}_a$

replaced by $s\,\mathbf{u}_a$. The a priori bounds, as linear functionals of $\mathbf{u}_a$ are also multiplied by s, and hence all solutions of $\mathscr{H}(s)\,\mathbf{u}=0$ lie in the interior of a suitably chosen surrounding sphere S_1. The results of the present section are summarized by the following theorem.

Theorem 2.5.1. Solutions (steady states) exist for Eqs. (2.4.1), (2.4.2) under boundary conditions (2.4.3), (2.4.4) or (2.4.17), (2.4.18).

2.6. Preliminaries on the Uniqueness of Steady States

The remaining four sections will be concerned with problems of uniqueness, stability, and asymptotic behavior of the steady states. The results are in close analogy with those obtained in Section 1.9, especially in regard to the behavior of the system in the transport and reaction limited regimes.

In the present section we shall use the concept of the index to show that the number of steady states is odd and to formulate a preliminary uniqueness criterion. A steady state is a solution of Eqs. (2.4.14) – (2.4.16) or their equivalent

$$\xi = \mathbf{H}\,\xi \tag{2.6.1}$$

where the operator $\mathbf{H}$ has been defined by Eq. (2.5.16). According to Theorem A.5 of the Appendix, the calculation of the index is based on the eigenvalue problem

$$\phi = \lambda\,\mathbf{L}\phi \tag{2.6.2}$$

where $\mathbf{L}$ is the Fréchét derivative of the operator $\mathbf{H}$. The operator $\mathbf{L}$ is obtained by using Eq. (A.8) as

$$\mathbf{L}\phi = \int_0^{r_0} G(r, r')\,\mathbf{A}(\xi(r'))\,\phi(r')\,dr' \tag{2.6.3}$$

where the matrix $\mathbf{A}$ is defined by

$$A_{ij}(\xi) = \frac{\partial \tilde{f}_i}{\partial \xi_j} \tag{2.6.4}$$

and $G(r, r')$ has been defined by Eq. (2.5.13). Eq. (2.6.2) can be written in the equivalent form

$$\frac{1}{r^2}\frac{d}{dr}\left(r^2\frac{d\phi}{dr}\right) = -\lambda\,\mathbf{A}(\xi(r))\,\phi, \tag{2.6.5}$$

$$r=0: \quad \frac{d\phi}{dr} = 0, \tag{2.6.6}$$

$$r=r_0: \quad \phi = 0. \tag{2.6.7}$$

According to Theorem A.5 the index of a steady state $\xi(r)$ is $(-1)^\beta$, where β is the sum of multiplicities of the eigenvalues $\lambda \in (0, 1)$ of Eqs. (2.6.2), or (2.6.5)–(2.6.7).

In the previous section it was shown that the rotation of the vector field $\mathbf{I}-\mathbf{H}$ on a sphere surrounding all the steady states is $+1$. This rotation is equal to the sum of the indices of the steady states, therefore the number of the steady states is odd, since the indices take only values ± 1. These remarks prove the following theorem

Theorem 2.6.1. If none of the steady states is such that $\lambda = 1$ is an eigenvalue of Eqs. (2.6.5)–(2.6.7), the number of steady states is odd.

The exceptional case of $\lambda = 1$ being an eigenvalue of Eqs. (2.6.5)–(2.6.7) can be interpreted as in Section (1.8).

The concept of the index can be used for deriving uniqueness criteria. If, for example, each steady state has index $+1$, there is exactly one steady state. This observation can be applied in spite of the unavailability of the steady states by using their a priori properties, for instance the fact that they lie in the invariant manifold $\tilde{\Gamma}(u_0)$. A first result is the theorem

Theorem 2.6.2. If for each $\xi \in V(u_0)$, the problem (2.6.5)–(2.6.7) does not have any eigenvalues λ in the interval $[0, 1]$, then Eq. (2.6.1) has a unique solution.

Proof. Since there is no eigenvalue in $[0, 1]$, $\beta = 0$ and the index of each steady state is $+1$. The sum of the indices of the steady states is also $+1$, therefore there is exactly one steady state.

One application of Theorem 2.6.2 is illustrated by Example 2.6.1. Generally, however, Theorem 2.6.2 is very difficult to apply because the eigenvalues of Eqs. (2.6.5)–(2.6.7) cannot be computed explicitly, as the elements of the matrix $\mathbf{A}$ are unknown functions of r. More useful uniqueness criteria will be developed in Section 2.7.

Example 2.6.1. A problem treated for uniqueness by Luss and Amundson [30] is

$$-\mathcal{M}_1 + \mathcal{M}_2 = 0 \tag{2.6.8}$$

with rate

$$f(c_1, T) = k_1\, c_1\, e^{-\frac{E}{R_g T}} \tag{2.6.9}$$

This system can be described by the variable T alone

$$\frac{1}{r^2}\frac{d}{dr}\left(r^2 \frac{dT}{dr}\right) = \frac{\Delta H}{k_c} \tilde{f}(T), \tag{2.6.10}$$

$$r = 0: \quad \frac{dT}{dr} = 0, \tag{2.6.11}$$

$$r = r_0: \quad T = T_0 \tag{2.6.12}$$

where

$$\tilde{f}(T)=k_1\left[c_{10}+\frac{k_c}{D_1\,\Delta H}(T-T_0)\right]e^{-\frac{E}{R_gT}}. \qquad (2.6.13)$$

The index of a steady state $T(r)$ is calculated from the eigenvalue problem

$$\frac{1}{r^2}\frac{d}{dr}\left(r^2\frac{d\phi}{dr}\right)=\frac{\Delta H}{k_c}\tilde{f}'(T(r))\,\phi, \qquad (2.6.14)$$

$$r=0:\ \frac{d\phi}{dr}=0, \qquad (2.6.15)$$

$$r=r_0:\quad \phi=0 \qquad (2.6.16)$$

where prime denotes differentiation. If the reaction is endothermic, $\Delta H>0$ and T takes values in the interval $[T_*, T_0]$, where

$$T_*=\max\left\{0,\,T_0-\frac{D_1\,\Delta H}{k_c}c_{10}\right\}. \qquad (2.6.17)$$

In this interval $\Delta H\,\tilde{f}'(T)>0$ and the eigenvalues of the problem (2.6.14)–(2.6.16) are negative, therefore the steady state is unique. If the reaction is exothermic, $\Delta H<0$ and $T_0\leqq T(r)\leqq T^*$, where

$$T^*=T_0+\frac{D_1\,|\Delta H|}{k_c}c_{10}. \qquad (2.6.18)$$

In this case $\Delta H\,\tilde{f}'(T)\geqq 0$ for all $T\in[T_0, T^*]$, provided

$$\frac{E\,|\Delta H|\,D_1\,c_{10}}{R_g\,T_0^2\,k_c}\leqq 1. \qquad (2.6.19)$$

Thus, the inequality (2.6.19) is a sufficient condition for the uniqueness of the steady state. A simple result of (2.6.19) is that the steady state is unique for a sufficiently dilute gas mixture (small c_{10}) [30].

2.7. Reaction Limited Regime-Uniqueness

As in the case of a single reaction, Section 2.3, we shall show that when the reaction rates are small compared to the transport rates the steady state is unique. In this respect, the situation is analogous with that of the uniform open systems of Section 1.9.

Theorem 2.7.1 [16]. Eqs. (2.4.14)–(2.4.16) have a unique solution for sufficiently small values of the radius r_0.

Proof. The proof will make use of Theorem (2.6.2). By forming the scalar product of Eq. (2.6.5) with $r^2\phi$, integrating, and integrating by parts, there is obtained

$$\lambda = \frac{\int_0^{r_0} r^2 \left(\frac{d\phi}{dr}, \frac{d\phi}{dr} \right) dr}{\int_0^{r_0} r^2 (\phi, \mathbf{A}(\xi)\, \phi)\, dr} \tag{2.7.1}$$

where (ϕ, ψ) is the scalar product in E_R, i.e.

$$(\phi, \psi) = \sum_{i=1}^{R} \bar{\phi}_i \psi_i \tag{2.7.2}$$

and the bar denotes complex conjugates. We wish to obtain a lower bound on λ and to this end we transform the numerator using POINCARÉ'S inequality (see Lemma 2.7.1 below):

$$\int_0^{r_0} r^2 \left(\frac{d\phi}{dr}, \frac{d\phi}{dr} \right) dr \geqq \frac{6}{r_0^2} \int_0^{r_0} r^2 (\phi, \phi)\, dr. \tag{2.7.3}$$

Since only real λ are of interest, the denominator can be estimated by considering real ϕ only:

$$(\phi, \mathbf{A}\phi) = (\phi, \mathbf{A}^s \phi) \leqq \alpha(\xi)\, (\phi, \phi) \tag{2.7.4}$$

where

$$A_{ij}^s = \tfrac{1}{2} (A_{ij} + A_{ji}) \tag{2.7.5}$$

and $\alpha(\xi)$ is the largest positive eigenvalue of the matrix $\mathbf{A}^s(\xi)$. Furthermore by defining

$$\alpha_M = \max_{\xi \in \tilde{\Gamma}(u_0)} \alpha(\xi) \tag{2.7.6}$$

we obtain

$$\left| \int_0^{r_0} r^2 (\phi, \mathbf{A}\phi)\, dr \right| \leqq \alpha_M \int_0^{r_0} r^2 (\phi, \phi)\, dr. \tag{2.7.7}$$

The inequalities (2.7.3), (2.7.7) are introduced in Eq. (2.7.1) to give

$$|\lambda| > \frac{6}{\alpha_M r_0^2}. \tag{2.7.8}$$

Noting that $\mathbf{A}(\xi)$ and α_M do not depend on r_0 we have the following sufficient criterion for uniqueness:

$$r_0 \leqq \left(\frac{6}{\alpha_M} \right)^{\frac{1}{2}}. \tag{2.7.9}$$

Theorem 2.7.1 can be explained as follows. When r_0 is small, the nonlinear source term due to the chemical reaction is a small perturbation on the linear diffusion equation. The state variables are close to their surface values and the steady state is unique. It is possible that Eq. (2.7.9) implies that Eq. (2.6.1) is a contraction mapping and can be solved by the method of successive approximations.

Lemma 2.7.1. If ϕ is continuously differentiable and vanishes at $r=r_0$ then

$$\int_0^{r_0} r^2\,\phi(r)\,dr \leqq \frac{r_0^2}{6}\int_0^{r_0} r^2\left(\frac{d\phi}{dr}\right)^2 dr.$$

Proof. This is a special case of POINCARÉ'S inequality and can be proved as follows

$$\phi^2(r)=\left[-\int_r^{r_0}\phi'(x)\,dx\right]^2 \leqq \int_r^{r_0}\frac{dx}{x^2}\int_r^{r_0} x^2\,\phi'^2(x)\,dx$$

$$\leqq\left(\frac{1}{r}-\frac{1}{r_0}\right)\int_0^{r_0} x^2\,\phi'^2(x)\,dx$$

where the Cauchy-Schwartz inequality was applied to the product $(x\,\phi'(x))\,(1/x)$. Multiplication of the last inequality by r^2 and integration gives the desired result.

Example 2.7.1. This is a continuation of Example 2.6.1. In order to apply the uniqueness criterion (2.7.9) to an exothermic reaction we need to compute the maximum of $|\Delta H\, \bar{f}'(T)/k_c|$ in $T_0 \leqq T \leqq T^*$. This is

$$\alpha_M = k_1\left(\frac{E\,k_c\,|\Delta H|\,c_1(T_1)}{R_g\,T_1^2}-\frac{1}{D_1}\right)e^{-\frac{E}{R_g T_1}} \tag{2.7.10}$$

where

$$T_1=\begin{cases}\dfrac{T^*}{1+2\,\dfrac{R_g\,T^*}{E}}, & \dfrac{E}{R_g\,T_0}\geqq 2\,\dfrac{T^*}{T^*-T_0}\\[2ex] T_0, & \dfrac{E}{R_g\,T_0}\leqq 2\,\dfrac{T^*}{T^*-T_0}.\end{cases} \tag{2.7.11}$$

Eq. (2.7.10) has been derived in [30] and compared with some numerical results. It turns out that the quantity $(6/\alpha_M)^{\frac{1}{2}}$ is one to two times higher than the value of r_0 at which the number of steady states changes from one to three.

Example 2.7.2 [16]. This is a uniqueness analysis on the reaction system of Example 2.4.2. Both reactions will be taken as irreversible and exothermic with rates

$$f_1(c_1, c_3, T)=k_1\,c_1\,c_3\,e^{-\frac{E_1}{R_g T}}, \tag{2.7.12}$$

$$f_2(c_2, c_3, T)=k_2\,c_2\,c_3\,e^{-\frac{E_2}{R_g T}}. \tag{2.7.13}$$

Introducing the extents ξ_1, ξ_2 defined by Eqs. (2.4.45)–(2.4.50), the following boundary value problem is obtained:

$$\frac{1}{r^2}\frac{d}{dr}\left(r^2\frac{d\xi_j}{dr}\right) = -\tilde{f}_j(\xi_1, \xi_2) \qquad j=1,2, \tag{2.7.14}$$

$$r=0:\ \frac{d\xi_j}{dr}=0; \qquad r=r_0:\ \xi_j=0 \qquad j=1,2. \tag{2.7.15}$$

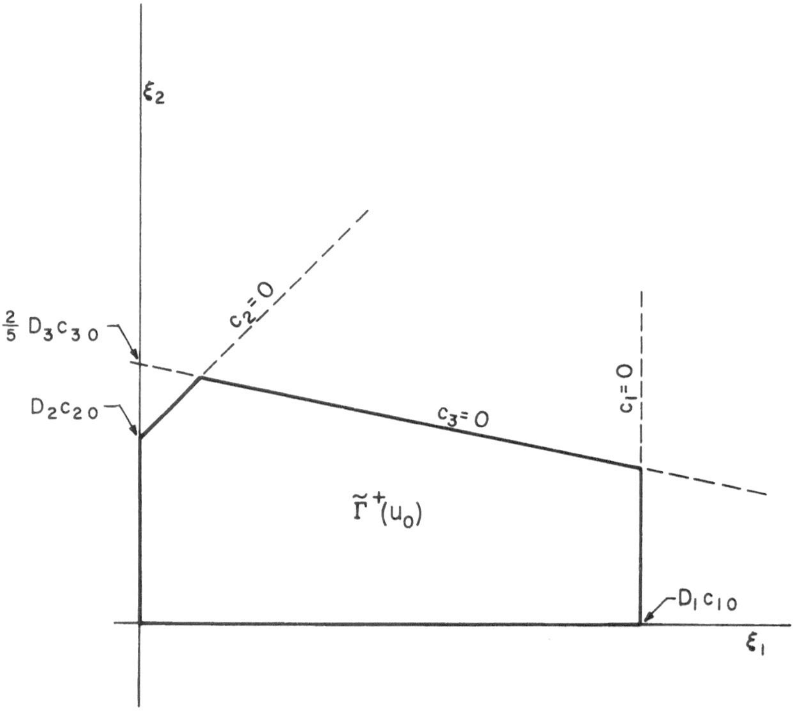

Fig. 2.7.1. The accessible region of an invariant manifold for irreversible reactions

From $\tilde{f}_1 \geqq 0, \tilde{f}_2 \geqq 0$ there follows $\xi_1 \geqq 0, \xi_2 \geqq 0$ and only the positive section of the invariant manifold of Fig. 2.4.2 is accessible. This is denoted by $\tilde{\Gamma}^+(u_0)$ and is shown in Fig. 2.7.1. The uniqueness eigenvalue problem is

$$\frac{1}{r^2}\frac{d}{dr}\left(r^2\frac{d\phi_1}{dr}\right) = -\lambda(A_{11}\phi_1 + A_{12}\phi_2), \tag{2.7.16}$$

$$\frac{1}{r^2}\frac{d}{dr}\left(r^2\frac{d\phi_2}{dr}\right) = -\lambda(A_{21}\phi_1 + A_{22}\phi_2), \tag{2.7.17}$$

$$r=0:\ \frac{d\phi_1}{dr}=\frac{d\phi_2}{dr}=0; \qquad r=r_0:\ \phi_1=\phi_2=0 \tag{2.7.18}$$

where

$$A_{11}=\left(\frac{E_1\,|\Delta H_1|\,c_1\,c_3}{k_c\,R_g\,T^2}-\frac{c_3}{D_1}-\frac{c_1}{D_3}\right)k_1\,e^{-\frac{E_1}{R_gT}}, \tag{2.7.19}$$

$$A_{12}=\left(\frac{E_1\,|\Delta H_2|\,c_1\,c_3}{k_c\,R_g\,T^2}-\frac{5c_1}{2D_3}\right)k_1\,e^{-\frac{E_1}{R_gT}}, \tag{2.7.20}$$

$$A_{21}=\left(\frac{E_2\,|\Delta H_1|\,c_2\,c_3}{k_c\,R_g\,T^2}+\frac{c_3}{D_2}-\frac{c_2}{2D_3}\right)k_2\,e^{-\frac{E_2}{R_gT}}, \tag{2.7.21}$$

$$A_{22}=\left(\frac{E_2\,|\Delta H_2|\,c_2\,c_3}{k_c\,R_g\,T^2}-\frac{c_3}{D_2}-\frac{5c_2}{2D_3}\right)k_2\,e^{-\frac{E_2}{R_gT}}. \tag{2.7.22}$$

In the present example it is difficult to obtain explicit uniqueness criteria in terms of the various parameters. Some progress can be made when the parameters are specified numerically. Let us choose the numerical values

$$\Delta H_1=-24{,}686\ \text{cal gmole}^{-1},$$
$$\Delta H_2=-316{,}192\ \text{cal gmole}^{-1},$$
$$k_1=6.75\times10^{9}\ \text{sec}^{-1}\,(\text{gmole cm}^{-3})^{-1},$$
$$k_2=7.35\times10^{-10}\ \text{sec}^{-1}\,(\text{gmole cm}^{-3})^{-1},$$
$$E_1=13{,}000\ \text{cal gmole}^{-1},$$
$$E_2=21{,}000\ \text{cal gmole}^{-1},$$
$$k_c=5.30\times10^{-4}\ \text{cal cm}^{-1}\,\text{sec}^{-1}\,{}^\circ\text{K}^{-1},$$
$$D_1=0.0100\ \text{cm}^2\,\text{sec}^{-1},$$
$$D_2=0.00798\ \text{cm}^2\,\text{sec}^{-1},$$
$$D_3=0.00935\ \text{cm}^2\,\text{sec}^{-1},$$
$$c_{10}=2.2\times10^{-5}\ \text{gmole cm}^{-3},$$
$$c_{20}=1.0\times10^{-5}\ \text{gmole cm}^{-3},$$
$$c_{30}=3.0\times10^{-5}\ \text{gmole cm}^{-3},$$
$$T_0=573\ {}^\circ\text{K}.$$

Using these values and the invariant manifold of Fig. 2.6.1, we obtain by a simple calculation:

$$A_{11}(\xi)<0,\qquad \xi\in\tilde{\Gamma}^{+}(u_0). \tag{2.7.23}$$

Hence, Eq. (2.7.1) yields

$$\lambda \geqq \frac{6}{r_0^2} \frac{\int_0^{r_0} r^2 (\phi_1^2 + \phi_2^2)\, dr}{\int_0^{r_0} r^2 [(A_{12} + A_{21})\, \phi_1\, \phi_2 + A_{22}\, \phi_2^2]\, dr}. \tag{2.7.24}$$

To estimate the denominator, we consider the matrix:

$$\begin{bmatrix} 0 & \dfrac{A_{12}+A_{21}}{2} \\ \dfrac{A_{12}+A_{21}}{2} & A_{22} \end{bmatrix} \tag{2.7.25}$$

and let $\alpha(\xi)$ be its largest eigenvalue:

$$\alpha(\xi) = \tfrac{1}{2}\{A_{22} + [A_{22}^2 + (A_{12} + A_{21})^2]^{\frac{1}{2}}\}. \tag{2.7.26}$$

The maximum of $\alpha(\xi)$ in $\tilde{\Gamma}^+(u_0)$ can be calculated only numerically. However, at the cost of obtaining weaker bounds we can use the inequality

$$\alpha_M < \tfrac{1}{2}\{a_{22} + [a_{22}^2 + (a_{12} + a_{21})^2]^{\frac{1}{2}}\} \tag{2.7.27}$$

where a_{ij} is the maximum of $|A_{ij}|$ in $\tilde{\Gamma}^+(u_0)$. With the help of Fig. 2.6.1 and a few calculations there is obtained

$$a_{12} < 65.72, \quad a_{21} < 6.67, \quad a_{22} < 0.40, \quad a_M < 36.40. \tag{2.7.28}$$

Hence, a sufficient condition for uniqueness is

$$r_0 < \left(\frac{6}{\alpha_M}\right)^{\frac{1}{2}} = 0.406\ \text{cm}. \tag{2.7.29}$$

Remarks

(i) Due to the fact that the uniqueness criterion (2.7.9) is only sufficient and to the additional approximations involved in its application to specific cases, conditions such as Eq. (2.7.29) are quite conservative.

(ii) Luss and Amundson [29], and Luss [31] have used bifurcation point theory to obtain sharper uniqueness criteria for the case of a single reaction. It would be very interesting to extend this promising approach to systems of chemical reactions.

(iii) Uniqueness can also be investigated without the use of topological methods, Luss and Amundson [30]. If ξ, ζ are two solutions of Eq. (2.6.1), then by subtraction and application of the mean value theorem there is obtained

$$\frac{1}{r^2}\frac{d}{dr}\left(r^2\frac{d\phi_j}{dr}\right) = -\sum_{k=1}^{R} A_{jk}(\xi^{(j)})\,\phi_k \qquad j=1,\dots,R \qquad (2.7.30)$$

where $\phi_j = \zeta_j - \xi_j$ and $\xi_i \leqq \xi_i^{(j)} \leqq \zeta_i$ for $i,j=1,\dots,R$. If the eigenvalue problem (2.7.30), (2.6.6), (2.6.7) has no nontrivial solution for all sets of functions $\xi^{(1)},\dots,\xi^{(R)}$ in $\tilde{\Gamma}(u_0)$, then the steady state is unique. The two eigenvalue problems (2.6.5)–(2.6.7) and (2.7.30), (2.6.6), (2.6.7) lead to the same uniqueness conditions when applied to concrete examples.

2.8. Asymptotic Behavior in the Transport Limited Regime

In this section we shall generalize some results of Section 2.3 concerning the asymptotic behavior of distributed chemical systems. We shall show that as the size of the system increases, the rates of chemical reactions and transport processes are appreciable only in a surface layer of constant thickness while the interior of the system is very nearly at equilibrium. Using this result, we shall obtain some a priori bounds for the effectiveness factor. The related problem of uniqueness of the steady state at large values of r_0 has not been solved rigorously. It is expected, however, that as in the case of a single reaction, the steady state is unique for sufficiently large r_0.

Bounds for Surface Fluxes and Effectiveness Factors [17]

An important property of distributed chemical systems is that in addition to the a priori bounds for the state variables, a priori bounds can be obtained for the derivatives of the state variables, i.e., the fluxes. The procedure of obtaining the a priori bounds for the fluxes is based on the following observation: Regardless of the value of the radius r_0, the a priori bounds for the concentrations will be violated if the values of the fluxes become too high.

Let us use the notation

$$c_{im} = \min c_i, \qquad c_{iM} = \max c_i, \qquad (2.8.1)$$

$$T_m = \min T, \qquad T_M = \max T, \qquad (2.8.2)$$

$$\xi_{jm} = \min \xi_j, \qquad \xi_{jM} = \max \xi_j, \qquad (2.8.3)$$

$$f_{jM} = \max |f_j|, \qquad F_{iM} = \max |F_i|, \qquad (2.8.4)$$

$$F_i = \sum_{j=1}^{R} \nu_{ij} f_j, \qquad F_{N+1} = -\sum_{j=1}^{R} \Delta H_j f_j \qquad (2.8.5)$$

where all the minima and maxima are evaluated over the region $\tilde{\Gamma}(u_0)$. By integrating Eq.(2.4.14) there is obtained

$$\xi_j'(r)=\xi_j'(r_0)+\int_r^{r_0}\tilde{f}_j(\boldsymbol{\xi}(y))\,dy+2\int_r^{r_0}\frac{\xi_j'(y)}{y}\,dy \tag{2.8.6}$$

where the prime denotes derivative. Integration of Eq.(2.8.6) yields

$$\xi_j(r)=-\xi_j'(r_0)(r_0-r)-\int_r^{r_0}\left[\int_y^{r_0}\tilde{f}_j\,dz\right]dy-2\int_r^{r_0}\left[\int_y^{r_0}\frac{\xi_j'(z)}{z}\,dz\right]dy. \tag{2.8.7}$$

If $\xi_j'(r_0)\geqq 0$, Eq.(2.8.6) gives $\xi_j'(r)\geqq 0$ for all $r\in[r_1, r_0]$, where

$$r_1=r_0-\frac{|\xi_j'(r_0)|}{f_{jM}}. \tag{2.8.8}$$

By setting $r=r_1$ in Eq.(2.8.7) there is obtained:

$$\xi_j'(r_0)\leqq(-2\xi_{jm}f_{jM})^{1/2}. \tag{2.8.9}$$

If $\xi_j'(r_0)\leqq 0$, Eq.(2.8.6) gives $\xi_j'(r)\leqq 0$ for all $r\in[r_1, r_0]$ and by setting $r=r_1$ in Eq.(2.8.7) there is obtained

$$\xi_j'(r_0)\geqq-(2\xi_{jM}f_{jM})^{1/2}. \tag{2.8.10}$$

These results can be combined to

$$-(2\xi_{jM}f_{jM})^{1/2}\leqq\xi_j'(r_0)\leqq(-2\xi_{jm}f_{jM})^{1/2}. \tag{2.8.11}$$

Bounds for the derivatives of the concentrations can be obtained in a similar way

$$\left[\frac{2F_{iM}(c_{iM}-c_{i0})}{D_i}\right]^{1/2}\leqq c_i'(r_0)\leqq\left[\frac{2F_{iM}(c_{i0}-c_{im})}{D_i}\right]^{1/2}, \tag{2.8.12}$$

$$\left[\frac{2F_{N+1,M}(T_M-T_0)}{k_c}\right]\leqq T'(r_0)\leqq\left[\frac{2F_{N+1,M}(T_0-T_m)}{k_c}\right]^{1/2}. \tag{2.8.13}$$

The same exactly bounds hold in the case of planar regions.

Effectiveness factors can be defined for each of the reactions or each of the species. Using the definition of Section 2.3, we can write the effectiveness factor for the ith species as

$$\mathrm{E}_i^{(3)}=-\frac{3D_i\,c_i'(r_0)}{r_0\,F_i(\mathbf{u}_0)} \tag{2.8.14}$$

and for the jth reaction as

$$\tilde{\mathrm{E}}_j^{(3)}=-\frac{3\,\xi_j'(r_0)}{r_0\,f_j(\mathbf{u}_0)}. \tag{2.8.15}$$

If $c_i'(r_0)>0$ and $F_i(\mathbf{u}_0)<0$ we can use Eq.(2.8.12) to obtain

$$\mathrm{E}_i^{(3)} \leqq \frac{3}{h_i}\left[\frac{2\,F_{iM}(c_{i0}-c_{im})}{F_i(\mathbf{u}_0)\,c_{i0}}\right]^{\frac{1}{2}} \tag{2.8.16}$$

where

$$h_i = r_0\left(\frac{F_i(\mathbf{u}_0)}{c_{i0}\,D_i}\right)^{\frac{1}{2}}. \tag{2.8.17}$$

Similarly, if $\xi_j(r_0)<0$ and $f_j(\mathbf{u}_0)>0$ we obtain

$$\tilde{\mathrm{E}}_j^{(3)} \leqq \frac{3}{\tilde{h}_j}\left(\frac{2 f_{jM}}{f_j(\mathbf{u}_0)}\right)^{\frac{1}{2}} \tag{2.8.18}$$

where

$$\tilde{h}_j = r_0\left(\frac{f_j(\mathbf{u}_0)}{|\xi_{jm}|}\right)^{\frac{1}{2}}. \tag{2.8.19}$$

Lower bounds for $\mathrm{E}_i^{(3)}$, $\tilde{\mathrm{E}}_j^{(3)}$, can be similarly obtained if these quantities are negative. Note that the quantities multiplying $1/h_i$, $1/\tilde{h}_j$ in Eqs.(2.8.16), (2.8.18) are independent of r_0, and independent of any common multiplicative factor in the reaction rates. Thus, the various effectiveness factors depend on the appropriate Thiele moduli in the same way as in the case of a single reaction.

The bounds in Eqs.(2.8.11)–(2.8.13) are general, but rather conservative. For example, in the case of a single irreversible isothermal reaction with rate $k\,c_1^n$, Eq.(2.8.16) gives a bound higher than the actual value of $\mathrm{E}_i^{(3)}$ by a factor of $(n+1)^{\frac{1}{2}}$. In more complicated situations the bounds will be even more conservative. Sharper bounds can be obtained in special cases by a technique illustrated in Example 2.8.2.

The Equilibrium State in Distributed Systems

The equilibrium state $\mathbf{u}^*$ in a given manifold $\Gamma(u_0)$ is the intersection of $\Gamma(u_0)$ with the kinetic equilibrium manifold Ω', defined by Eq.(1.5.4). In terms of the $\boldsymbol{\xi}$ variables, the equilibrium state $\boldsymbol{\xi}^*$ is the solution of $\tilde{f}(\boldsymbol{\xi})=0$. Obviously, the equilibrium state in distributed systems depends on the transport properties. In some cases of empirical kinetics, such as in Example 2.8.2, there may exist more than one equilibrium states.

We wish now to investigate the way in which an open distributed system can approach the equilibrium state. To this end we integrate Eqs.(2.4.14)–(2.4.16) and use the inequalities (2.8.9) and (2.8.10) to obtain

$$-r_0^2(2\,\xi_{jM}\,f_{jM})^{\frac{1}{2}} \leqq \int_0^{r_0} r^2 f_j\,dr \leqq r_0^2(-2\,\xi_{jm}\,f_{jM})^{\frac{1}{2}}. \tag{2.8.20}$$

If the jth reaction is irreversible, f_j has a fixed sign and vanishes only at equilibrium, it follows then from Eq.(2.4.14)–(2.4.16) that ξ_j is monotonic. As r_0 increases, the bounds of the integral in Eq.(2.8.20) increase as r_0^2, but the volume of integration increases as r_0^3. Thus, f_j must be negligible except in a surface layer with thickness $(2\,|\xi_{jm}|\,f_{jM}^{-1})^{\frac{1}{2}}$. The reaction takes place in this surface layer and the interior is very near equilibrium.

In the case of a reversible reaction, f_j has no fixed sign, ξ_j need not be monotonic, and sustained oscillations cannot be ruled out except by introducing entropy considerations. Suppose that there exists a positive definite form $\sigma(\mathbf{u})$, vanishing only on the equilibrium manifold and satisfying a conservation equation of the type

$$\sigma = \operatorname{div} \mathbf{J}_s \tag{2.8.21}$$

where

$$\mathbf{J}_s = \sum_{i=1}^{N} a_i(\mathbf{u}) \operatorname{grad} c_i + a_{N+1}(\mathbf{u}) \operatorname{grad} T \tag{2.8.22}$$

and the coefficients a_i are continuous functions of the state $\mathbf{u}$. By integrating Eq.(2.8.21) over the spherical region, there is obtained

$$\int_0^{r_0} r^2 \sigma\, dr = r_0^2 \sum_{i=1}^{N} a_i(\mathbf{u}_0)\, c_i'(r_0) + r_0^2\, a_{N+1}(\mathbf{u}_0)\, T'(r_0) \tag{2.8.23}$$

where the prime denotes derivative. By introducing Eq.(2.8.12) in (2.8.23) there is obtained

$$\int_0^{r_0} r^2 \sigma\, dr \leqq r_0^2\, b \tag{2.8.24}$$

where the constant b is independent of r_0. This inequality implies that the state of the system is appreciably removed from equilibrium only in a fraction $3b/r_0\,\sigma(\mathbf{u}_0)$ of the total volume. Intuitively, it is expected that the reactive region lies entirely near the surface. However, this has not been proved yet in general.

If the transport-kinetic model were compatible with thermodynamics, the entropy production would be a suitable positive definite function. In this case, however, it would be a very tedious task to establish bounds of the type (2.8.11)–(2.8.13). In the case of the simplified transport model, Eqs.(2.4.1) and (2.4.2), a positive definite quantity σ satisfying Eqs.(2.8.21) to (2.8.22) cannot be constructed for an arbitrary kinetic model. In many special cases, the construction is possible although σ need not have thermodynamic significance. This will be illustrated in the following examples [17].

Example 2.8.1. Reversible Isomerization Reactions. The three isomeric forms of a compound are interconverted in a porous catalyst according to the mechanism

$$\mathcal{M}_1 + \mathcal{X} = \mathcal{Y}, \tag{2.8.25}$$

$$\mathcal{M}_2 + \mathcal{X} = \mathcal{Y}, \tag{2.8.26}$$

$$\mathcal{M}_3 + \mathcal{X} = \mathcal{Y} \tag{2.8.27}$$

where $\mathcal{X}$ and $\mathcal{Y}$ are the unoccupied and occupied catalyst cites. We shall consider isothermal conditions, assuming that the enthalpies of the three forms are very nearly equal. The net rates of the three reactions are taken as

$$k_1^+ c_x c_1 - k_1^- c_y; \quad k_2^+ c_x c_2 - k_2^- c_y; \quad k_3^+ c_x c_3 - k_3^- c_y \tag{2.8.28}$$

where c_1, c_2, c_3, c_x, c_y are the concentrations of $\mathcal{M}_1, \mathcal{M}_2, \mathcal{M}_3, \mathcal{X}, \mathcal{Y}$. Since $\mathcal{X}$ and $\mathcal{Y}$ do not diffuse

$$-(k_1^+ c_1 + k_2^+ c_2 + k_3^+ c_3) c_x + (k_1^- + k_2^- + k_3^-) c_y = 0, \tag{2.8.29}$$

$$c_x + c_y = c_0 \tag{2.8.30}$$

where c_0 is the total concentration of catalyst cites. Eqs. (2.8.29) and (2.8.30) yield:

$$c_x = \frac{c_0}{1+\alpha}, \tag{2.8.31}$$

$$c_y = \frac{\alpha c_0}{1+\alpha} \tag{2.8.32}$$

where

$$\alpha = \frac{k_1^+ c_1 + k_2^+ c_2 + k_3^+ c_3}{k_1^- + k_2^- + k_3^-}. \tag{2.8.33}$$

From the reactions (2.8.25)–(2.8.27) one obtains the two independent reactions

$$\mathcal{M}_2 - \mathcal{M}_1 = 0, \tag{2.8.34}$$

$$\mathcal{M}_3 - \mathcal{M}_1 = 0 \tag{2.8.35}$$

with rates

$$f_1 = \frac{c_0}{1+\alpha} (k_2^+ c_2 - k_2^- \alpha), \tag{2.8.36}$$

$$f_2 = \frac{c_0}{1+\alpha} (k_3^+ c_3 - k_3^- \alpha). \tag{2.8.37}$$

In the 3-dimensional space of c_1, c_2, c_3, the equilibrium manifold, defined by $f_1 = f_2 = 0$, is the straight line

$$c_i = \frac{k_i^-}{k_i^+} s \qquad i = 1, 2, 3 \tag{2.8.38}$$

where s is a parameter. The invariant manifold is the plane

$$D_1 c_1 + D_2 c_2 + D_3 c_3 = D_1 c_{10} + D_2 c_{20} + D_3 c_{30}. \tag{2.8.39}$$

The equilibrium state corresponding to given surface conditions c_{10}, c_{20}, c_{30} is obtained from Eqs. (2.8.36), (2.8.37) as

$$c_i^* = \frac{k_i^-}{k_i^+} \alpha^* \tag{2.8.40}$$

where

$$\alpha^* = \frac{D_1 c_{10} + D_2 c_{20} + D_3 c_{30}}{D_1 \frac{k_1^-}{k_1^+} + D_2 \frac{k_2^-}{k_2^+} + D_3 \frac{k_3^-}{k_3^+}}. \tag{2.8.41}$$

Extents ξ_1, ξ_2 can be introduced with the equilibrium point as a reference:

$$D_1(c_1 - c_1^*) = -(\xi_1 + \xi_2), \tag{2.8.42}$$

$$D_2(c_2 - c_2^*) = \xi_1, \tag{2.8.43}$$

$$D_3(c_3 - c_3^*) = \xi_2 \tag{2.8.44}$$

so that $\xi_1^* = \xi_2^* = 0$. The conservation equations are

$$\frac{1}{r^2} \frac{d}{dr}\left(r^2 \frac{d\xi_1}{dr}\right) = \frac{c_0}{(1+\alpha)(k_1^- + k_2^- + k_3^-)} (f_{11} \xi_1 + f_{12} \xi_2), \tag{2.8.45}$$

$$\frac{1}{r^2} \frac{d}{dr}\left(r^2 \frac{d\xi_2}{dr}\right) = \frac{c_0}{(1+\alpha)(k_1^- + k_2^- + k_3^-)} (f_{21} \xi_1 + f_{22} \xi_2) \tag{2.8.46}$$

where

$$f_{11} = \frac{k_1^+ k_2^-}{D_1} + \frac{k_2^+ k_1^-}{D_2} + \frac{k_2^+ k_3^-}{D_2}, \tag{2.8.47}$$

$$f_{12} = k_2^- \left(\frac{k_2^+}{D_2} - \frac{k_1^+}{D_1}\right), \tag{2.8.48}$$

$$f_{21} = k_3^- \left(\frac{k_2^+}{D_2} - \frac{k_1^+}{D_1}\right), \tag{2.8.49}$$

$$f_{22} = -\left(\frac{k_3^+ k_1^-}{D_3} + \frac{k_3^+ k_2^-}{D_3} + \frac{k_1^+ k_3^-}{D_1}\right). \tag{2.8.50}$$

In the present example thermodynamics can be used as a guide for the construction of a positive definite form σ. For purely diffusive transport, we may define

$$J_s = \sum_{i=1}^{3} \frac{\mu_i}{T} D_i \frac{dc_i}{dr} \tag{2.8.51}$$

and obtain σ from Eq. (2.8.21) as

$$\sigma = \sum_{i=1}^{3} D_i \frac{dc_i}{dr} \frac{d}{dr}\left(\frac{\mu_i}{T}\right) + \frac{1}{T}(A_1 f_1 + A_2 f_2). \tag{2.8.52}$$

The quantities μ_i can be defined as chemical potentials of ideal gases

$$\mu_i = \mu_i^0 + R_g T \ln \frac{c_i}{c} \qquad i = 1, 2, 3, \tag{2.8.53}$$

$$c = c_1 + c_2 + c_3 \tag{2.8.54}$$

and the quantities μ_i^0, which are functions of the temperature only, are chosen to make $A_1 f_1 + A_2 f_2$ positive definite, for example

$$\mu_1^0 = 0, \quad \mu_2^0 = R_g T \ln \frac{k_1^- k_2^+}{k_2^- k_1^+}, \quad \mu_3^0 = R_g T \ln \frac{k_1^- k_3^+}{k_1^+ k_3^-}. \tag{2.8.55}$$

It is obvious that the kinetic and equilibrium manifolds coincide. By introducing the expressions for μ_i in Eq. (2.8.51) there is obtained

$$\sigma = R_g \sum_{i=1}^{3} \frac{D_i}{c_i} \left(\frac{dc_i}{dr}\right)^2 + \frac{c_0 R_g}{1+\alpha} \sum_{i=1}^{3} (c_i k_i^+ - \alpha k_i^-) \ln \frac{k_i^+ c_i}{k_i^- \alpha} \tag{2.8.56}$$

where each term at the right is individually positive definite. If the same construction is attempted for a nonisothermal system, the term representing the contribution of transport to the entropy production is more complicated and is not necessarily positive definite due to the simplified nature of the transport model. In such a case it may be possible to construct a different type of σ, as illustrated in the next example [17].

Example 2.8.2. Two Irreversible Exothermic Reactions. This will be a continuation of the analysis on the physical system treated in Examples 2.4.2, and 2.7.2. In contrast with Example 2.8.1, the rate expressions (2.7.12) and (2.7.13) are not consistent with thermodynamics. Indeed, the kinetic equilibrium manifold $f_1 = f_2 = 0$ consisting of two disjoint pieces

$$\Omega_1': \; c_1 = 0 \text{ and } c_2 = 0, \tag{2.8.57}$$

$$\Omega_2': \; c_3 = 0 \tag{2.8.58}$$

cannot be identical with the thermodynamic manifold $A_1 = A_2 = 0$.

Fig. 2.8.1 shows the two pieces of the equilibrium set, the point Ω_1', and the line segment Ω_2', in two relative configurations. The functions

$$\begin{aligned}\Phi_1 &= (\xi_1 - D_1 c_{10})^2 + (\xi_2 - D_1 c_{10} - D_2 c_{20})^2 \\ &= D_1^2 c_1^2 + (D_1 c_1 + D_2 c_2)^2,\end{aligned} \tag{2.8.59}$$

$$\Phi_2 = \tfrac{1}{26}(2 D_3 c_{30} - 5\xi_2 - \xi_1)^2 = \tfrac{4}{26} D_3^2 c_3^2 \tag{2.8.60}$$

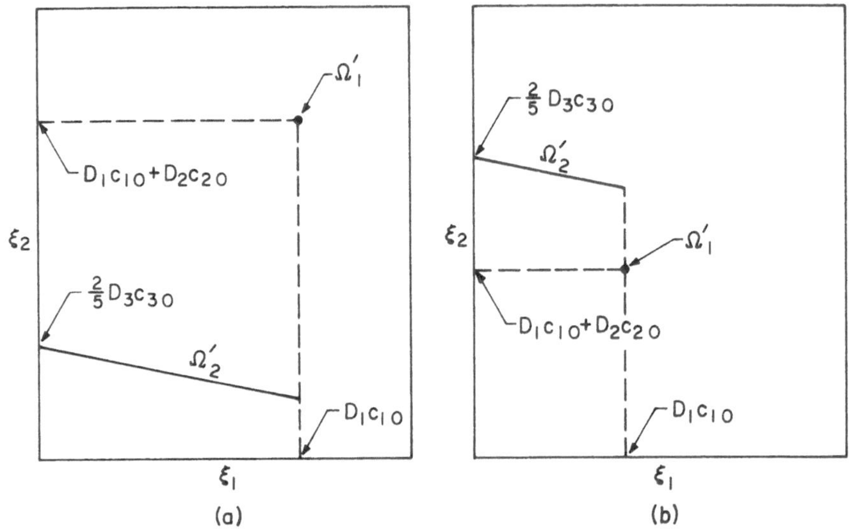

Fig. 2.8.1. The equilibrium manifold in Example 2.8.2

are the squares of the distances of a point (ξ_1, ξ_2) from Ω_1' and Ω_2'. Eqs. (2.4.14)–(2.4.16) show that ξ_1 and ξ_2 are monotonically decreasing. In the case of Fig. 2.8.1 a, only the line segment Ω_2' can be approached asymptotically as r_0 becomes large; Ω_1' cannot be approached without a crossing of the segment Ω_2' and a consequent increase in Φ_2. In the case of Fig. 2.8.1 b, Ω_1' is the only accessible equilibrium point. In Fig. 2.8.1 a, the point of approach depends not only on the invariant manifold, but on the reaction rates as well. A priori determination of the equilibrium point is possible only in the case of Fig. 2.8.1 b.

In order to show the asymptotic approach to equilibrium, we have to modify the general method based on Eq. (2.8.20) because the rates vanish at more than one point of the invariant manifold. Actually, we shall adopt a direct approach based on the behavior of the functions Φ_1, Φ_2. The proof will be given for the configuration of Fig. 2.8.1 a and other cases can be treated similarly. By direct computation we obtain:

$$\frac{1}{r^2}\frac{d}{dr}\left(r^2\frac{d\Phi_2}{dr}\right)=\frac{2}{26}\left(\frac{d\xi_1}{dr}+5\frac{d\xi_2}{dr}\right)^2+\frac{1}{D_3}g(\xi_1,\xi_2)\,\Phi_2 \tag{2.8.61}$$

where

$$g(\xi_1,\xi_2)=k_1\,c_1\,e^{-\frac{E_1}{R_gT}}+5k_2\,c_2\,e^{-\frac{E_2}{R_gT}}. \tag{2.8.62}$$

The quantity g vanishes at Ω_1' and is positive everywhere else. In the region below the segment Ω_2', $g(\xi_1,\xi_2)$ is bounded away from zero, i.e.,

$$g(\xi_1,\xi_2)\geqq\min\{g_1,g_2\}=g_m>0 \tag{2.8.63}$$

where

$$g_1=k_1\,c_{10}\,e^{-\frac{E_1}{R_gT_0}}+k_2\left(5c_{20}-\frac{2D_3}{D_2}c_{30}\right)e^{-\frac{E_2}{R_gT_0}}, \tag{2.8.64}$$

$$g_2=k_2\left[5c_{20}+\frac{1}{D_2}(6D_1\,c_{10}-2D_3\,c_{30})\right]e^{-\frac{E_2}{R_gT_0}}. \tag{2.8.65}$$

We then have

$$\frac{1}{r^2}\frac{d}{dr}\left(r^2\frac{d\Phi_2}{dr}\right)\geqq\frac{g_m}{D_3}\Phi_2, \tag{2.8.66}$$

$$r=0:\quad\frac{d\Phi_2}{dr}=0, \tag{2.8.67}$$

$$r=r_0:\quad\Phi_2=\tfrac{4}{26}D_3^2\,c_{30}^2 \tag{2.8.68}$$

whereby

$$\Phi_2(r)\leqq\frac{2r_0\,D_3^2\,c_{30}^2}{13\,r}\;\frac{\sinh\left(\frac{g_m\,r}{D_3}\right)^{\frac{1}{2}}}{\sinh\left(\frac{g_m\,r_0}{D_3}\right)^{\frac{1}{2}}} \tag{2.8.69}$$

which implies that the state (ξ_1,ξ_2) tends rapidly to equilibrium as r decreases. In fact, Φ_2 drops to less than 1% of its boundary value within a distance of about $5(D_3/g_m)^{\frac{1}{2}}$ from the surface.

Fig. 2.8.2 shows numerical solutions of Eqs. (2.7.14) and (2.7.15) for the parameter values given in Example 2.7.2 and for $r_0=0.1$ cm, 0.5 cm, 1.0 cm. Each curve is a phase plane plot of a solution $(\xi_1(r),\xi_2(r))$. The curves start at $r=0$ at varying distances from the equilibrium segment Ω_2' and reach the origin at $r=r_0$. Since $f_1\gg f_2$ the equilibrium point (ξ_1^*,ξ_2^*) lies very close to B. If k_2 is replaced by $k_2'=10k_2$ the equilibrium point is shifted towards A. The actual location of (ξ_1^*,ξ_2^*) on Ω_2' cannot be determined a priori; it depends on $\tilde{f}_1,\tilde{f}_2$.

Finally, we come to the subject of estimating the effectiveness factors. As has been noted before, the estimates given by Eqs. (2.8.18) are generally too high. We shall describe now a more accurate estimating technique,

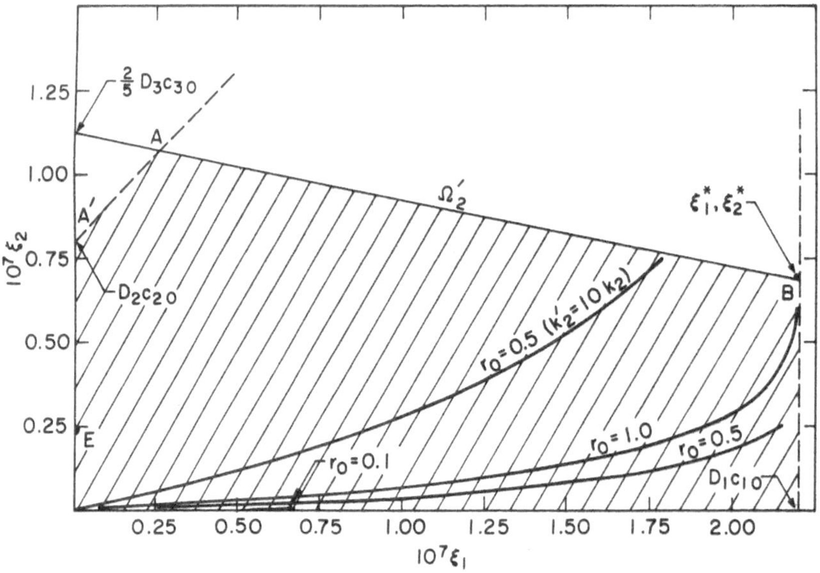

Fig. 2.8.2. Approach to equilibrium in the manifold of Example 2.8.2

applicable to irreversible reactions. Consider the two boundary value problems

$$D_1 \frac{1}{r^2} \frac{d}{dr}\left(r^2 \frac{dc_1}{dr}\right) = k_1\, c_1\, c_3\, e^{-\frac{E_1}{R_g T}}, \tag{2.8.70}$$

$$r=0:\ \frac{dc_1}{dr}=0; \qquad r=r_0:\ c_1=c_{10} \tag{2.8.71}$$

and

$$D_1 \frac{1}{r^2} \frac{d}{dr}\left(r^2 \frac{dv_1}{dr}\right) = k_1\, M_1\, v_1, \tag{2.8.72}$$

$$r=0:\ \frac{dv_1}{dr}=0; \qquad r=r_0:\ v_1=c_{10} \tag{2.8.73}$$

where

$$M_1 = \max_{\xi \in \tilde{\Gamma}^+} \left[c_3\, e^{-\frac{E_1}{R_g T}}\right]. \tag{2.8.74}$$

This maximum is attained at the point E of Fig. 2.8.2. In the problem (2.8.70) – (2.8.71), c_1 is the unknown function and c_3, T are considered as known functions of r; they are some solution of Eqs. (2.7.14) and

(2.7.15). The problem (2.8.72) and (2.8.73) is linear in the unknown v_1. By subtracting Eq. (2.8.72) from (2.8.70) there is obtained

$$D_1 \frac{1}{r^2} \frac{d}{dr}\left(r^2 \frac{d(c_1 - v_1)}{dr}\right) = k_1 c_1 [c_3 e^{-\frac{E_1}{R_g T}} - M_1] + k_1 M_1 (c_1 - v_1), \tag{2.8.75}$$

$$r=0: \frac{d(c_1 - v_1)}{dr} = 0; \quad r=r_0: c_1 - v_1 = 0 \tag{2.8.76}$$

whence $c_1 - v_1$ cannot have a negative minimum, i.e.,

$$c_1(r) - v_1(r) \geqq 0, \quad 0 \leqq r \leqq r_0, \tag{2.8.77}$$

$$r = r_0: \frac{dc_1}{dr} \leqq \frac{dv_1}{dr}. \tag{2.8.78}$$

Now dv_1/dr is easily obtained from Eqs. (2.8.72) and (2.8.73) as

$$\left(\frac{dv_1}{dr}\right)_{r_0} = \frac{c_{10}}{r_0} [r_0 \lambda \coth(\lambda r_0) - 1] \tag{2.8.79}$$

where

$$\lambda = \left(\frac{k_1 M_1}{D_1}\right)^{\frac{1}{2}}. \tag{2.8.80}$$

Finally then,

$$\mathrm{E}_1^{(3)} \leqq \frac{3(\lambda r_0 \coth(\lambda r_0) - 1)}{r_0^2 k_1 c_{30} e^{-\frac{E_1}{R_g T_0}}}. \tag{2.8.81}$$

Table 2.8.1 compares the values of effectiveness factors estimated by Eq. (2.8.81) with those obtained from an exact numerical solution. The agreement is good, especially at large r_0.

The effectiveness factors $\mathrm{E}_3^{(3)}, \tilde{\mathrm{E}}_1^{(3)}, \tilde{\mathrm{E}}_2^{(3)}$ can be estimated in a similar fashion since F_3, f_1, f_2 have fixed sign. The estimation of $\mathrm{E}_2^{(3)}$, on the other hand, is difficult because $F_2 = f_1 - f_2$ does not have fixed sign in $\tilde{\Gamma}^+(u_0)$. It would be interesting to use the given functional form of f_1, f_2 to further delimit the accessible section of $\tilde{\Gamma}(u_0)$ and thereby obtain useful estimates for $\mathrm{E}_2^{(3)}$.

Table 2.8.1. *Effectiveness factors in Example 2.8.2*

Effectiveness factor	$r_0 = 0.01$	$r_0 = 0.05$	$k_2' = 10 k_2$, $r_0 = 0.5$	$r_0 = 1.0$
From numerical solution	0.861	0.335	0.325	0.182
From Eq. (2.8.16)	2.84	0.568	0.568	0.284
From Eq. (2.8.81)	0.907	0.382	0.382	0.191

2.9. On the Stability of the Steady States

The problem of stability in distributed chemical reaction systems has been treated by AMUNDSON and RAYMOND [2], KUO and AMUNDSON [24–26], WEI [42], and others. These authors have obtained analytical stability (or instability) conditions in a few cases and have investigated other cases numerically. Generally, the analytical conditions are quite conservative due to the fact that the steady states cannot be obtained in explicit form and the analysis must be based entirely on the a priori bounds. An additional difficulty of the stability problem in chemical systems is the nonselfadjointness of the pertinent eigenvalue problem.

The present section is concerned with some stability results obtained by investigating the relationship between the index and the stability of a steady state. It turns out that steady states with indices -1 are unstable, therefore among the $2m+1$ steady states, at least m are unstable. Steady states with index $+1$, on the other hand, need not be stable. Furthermore, some of the uniqueness criteria of Section 2.7 turn out to be criteria for stability also. There is a close analogy between the results of the present section and those of Section 1.9. Since Section 1.9 is mathematically simpler, the reader may profit by reviewing it before proceeding to the present section.

In order to characterize a solution $\mathbf{u}(r)$ of Eqs. (2.4.1)–(2.4.4) relative to stability, it is necessary to consider the time dependent problem

$$\frac{\partial c_i}{\partial t}=\frac{D_i}{\varepsilon_p}\frac{1}{r^2}\frac{\partial}{\partial r}\left(r^2\frac{\partial c_i}{\partial r}\right)+\frac{F_i(\mathbf{u})}{\varepsilon_p}, \tag{2.9.1}$$

$$\frac{\partial T}{\partial t}=\frac{k_c}{\rho_p c_p}\frac{1}{r^2}\frac{\partial}{\partial r}\left(r^2\frac{\partial T}{\partial r}\right)+\frac{1}{\rho_p c_p}F_{N+1}(\mathbf{u}), \tag{2.9.2}$$

$$r=0:\ \frac{\partial c_i}{\partial r}=0,\qquad \frac{\partial T}{\partial r}=0, \tag{2.9.3}$$

$$r=r_0:\ c_i=c'_{i0},\qquad T=T'_0, \tag{2.9.4}$$

$$t=0:\ c_i=c'_i(r),\qquad T=T'(r) \tag{2.9.5}$$

obtained from Eqs. (2.1.31), (2.1.32) by assumming constant transport coefficients and setting

$$F_i=\sum_{j=1}^{R}\nu_{ij}f_j \qquad i=1,\dots,N, \tag{2.9.6}$$

$$F_{N+1}=-\sum_{j=1}^{R}\Delta H_j f_j. \tag{2.9.7}$$

The steady state $\mathbf{u}(r)$ will be called stable if given any $\varepsilon>0$ there exists a $\delta>0$ such that the solution $\mathbf{u}(t,r)$ of Eqs. (2.9.1)–(2.9.5) satisfies

$\|\mathbf{u}(t,r)-\mathbf{u}(r)\| \leqq \varepsilon$ for all $t>0$ provided $\|\mathbf{u}'(r)-\mathbf{u}(r)\| \leqq \delta$ and $|\mathbf{u}_0'-\mathbf{u}_0| \leqq \delta$. We shall actually consider the simpler and most likely equivalent problem of perturbations in the initial conditions only. Furthermore, we shall restrict the analysis to the linearized version of Eqs. (2.9.1), (2.9.2)

$$\frac{\partial \psi_i}{\partial t} = \frac{D_i}{\varepsilon_p} \frac{1}{r^2} \frac{\partial}{\partial r}\left(r^2 \frac{\partial \psi_i}{\partial r}\right) + \frac{1}{\varepsilon_p} \sum_{j=1}^{N+1} \frac{\partial F_i}{\partial u_j} \psi_j \qquad i=1,\dots,N, \tag{2.9.8}$$

$$\frac{\partial \psi_{N+1}}{\partial t} = \frac{k_c}{\rho_p c_p} \frac{1}{r^2} \frac{\partial}{\partial r}\left(r^2 \frac{\partial \psi_{N+1}}{\partial r}\right) + \frac{1}{\rho_p c_p} \sum_{j=1}^{N+1} \frac{\partial F_{N+1}}{\partial u_j} \psi_j, \tag{2.9.9}$$

$$r=0: \quad \frac{\partial \psi_i}{\partial r} = 0, \tag{2.9.10}$$

$$r=r_0: \quad \psi_i = 0, \tag{2.9.11}$$

$$t=0: \quad \psi_i = g_i(r) \tag{2.9.12}$$

where the derivatives $\partial F_i/\partial u_j$ are evaluated at the steady state under question and $g_i(r)$ is an arbitrary initial condition. The stability of the linearized problem is determined from the eigenvalue problem

$$\frac{D_i}{\varepsilon_p} \frac{1}{r^2} \frac{d}{dr}\left(r^2 \frac{d\psi_i}{dr}\right) + \frac{1}{\varepsilon_p} \sum_{j=1}^{N+1} \frac{\partial F_i}{\partial u_j} \psi_j = \mu \psi_i \qquad i=1,\dots,N, \tag{2.9.13}$$

$$\frac{k_c}{\rho_p c_p} \frac{1}{r^2} \frac{d}{dr}\left(r^2 \frac{d\psi_{N+1}}{dr}\right) + \frac{1}{\rho_p c_p} \sum_{j=1}^{N+1} \frac{\partial F_{N+1}}{\partial u_j} \psi_j = \mu \psi_{N+1}, \tag{2.9.14}$$

$$r=0: \quad \frac{d\psi_i}{dr} = 0 \qquad i=1,\dots,N+1, \tag{2.9.15}$$

$$r=r_0: \quad \psi_i = 0 \qquad i=1,\dots,N+1. \tag{2.9.16}$$

Now the relation between the stability of the nonlinear problem and its linearized version has not been firmly established as in the case of ordinary differential equations. Strictly speaking then, the term stability will refer to the linearized problem. A steady state $\mathbf{u}(r)$ will be called (locally) stable if all the eigenvalues μ of Eqs. (2.9.13)–(2.9.16) have negative real parts, and unstable if one or more eigenvalues μ have positive parts.

We wish to relate the eigenvalue problem (2.9.13)–(2.9.16) to the eigenvalue problem for uniqueness. Now, while the uniqueness eigenvalue problem refers to a time independent situation and can be described by the extents $\xi_1, \dots, \xi_R$, the stability eigenvalue problem (2.9.13)–(2.9.16) describes a time dependent situation and needs $N+1$ variables for its description. This can be verified by the fact that Eqs. (2.9.13)–(2.9.14) cannot be reduced to R equations. Consequently, in order to compare the stability and the uniqueness eigenvalue problems, we need to write the latter in terms of the $N+1$ variables, based directly on Eqs. (2.4.1)–(2.4.4):

$$D_i \frac{1}{r^2} \frac{d}{dr}\left(r^2 \frac{d\chi_i}{dr}\right) = -\lambda \sum_{j=1}^{N+1} \frac{\partial F_i}{\partial u_j} \chi_j \qquad i=1,\dots,N, \tag{2.9.17}$$

$$k_c \frac{1}{r^2} \frac{d}{dr}\left(r^2 \frac{d\chi_{N+1}}{dr}\right) = -\lambda \sum_{j=1}^{N+1} \frac{\partial F_{N+1}}{\partial u_j} \chi_j, \tag{2.9.18}$$

$$r=0:\ \frac{d\chi_i}{dr}=0; \qquad r=r_0:\ \chi_i=0. \tag{2.9.19}$$

Next, we shall define matrices

$$\mathbf{D}=\begin{bmatrix} D_1 & 0 & \dots & 0 & 0 \\ 0 & D_2 & \dots & 0 & 0 \\ \cdot & \cdot & \cdot & \cdot & \cdot \\ 0 & 0 & \dots & D_N & 0 \\ 0 & 0 & \dots & 0 & k_c \end{bmatrix}, \quad \mathbf{B}=\begin{bmatrix} \frac{\partial F_1}{\partial c_1} & \dots & \frac{\partial F_1}{\partial c_N} & \frac{\partial F_1}{\partial T} \\ \frac{\partial F_2}{\partial c_1} & \dots & \frac{\partial F_2}{\partial c_N} & \frac{\partial F_2}{\partial T} \\ \cdot & \cdot & \cdot & \cdot \\ \frac{\partial F_{N+1}}{\partial c_1} & \dots & \frac{\partial F_{N+1}}{\partial c_N} & \frac{\partial F_{N+1}}{\partial T} \end{bmatrix}, \tag{2.9.20}$$

$$\mathbf{P}=\begin{bmatrix} \varepsilon_p & 0 & \dots & 0 & 0 \\ 0 & \varepsilon_p & \dots & 0 & 0 \\ \cdot & \cdot & \cdot & \cdot & \cdot \\ 0 & 0 & \dots & \varepsilon_p & 0 \\ 0 & 0 & \dots & 0 & \rho_p c_p \end{bmatrix} \tag{2.9.21}$$

and rewrite the eigenvalue problems (2.9.13)–(2.9.16) and (2.9.17)–(2.9.19) as

$$\frac{1}{r^2} \frac{d}{dr}\left(r^2 \frac{d\boldsymbol{\psi}}{dr}\right) + \mathbf{D}^{-1}\mathbf{B}\boldsymbol{\psi} = \mu\,\mathbf{D}^{-1}\mathbf{P}\boldsymbol{\psi}, \tag{2.9.22}$$

$$r=0:\ \frac{d\boldsymbol{\psi}}{dr}=0; \qquad r=r_0:\ \boldsymbol{\psi}=0, \tag{2.9.23}$$

$$\frac{1}{r^2} \frac{d}{dr}\left(r^2 \frac{d\boldsymbol{\chi}}{dr}\right) = -\lambda\,\mathbf{D}^{-1}\mathbf{B}\boldsymbol{\chi}, \tag{2.9.24}$$

$$r=0:\ \frac{d\boldsymbol{\chi}}{dr}=0; \qquad r=r_0:\ \boldsymbol{\chi}=0. \tag{2.9.25}$$

It is expected that the two forms of the uniqueness eigenvalue problem (2.9.24)–(2.9.25) and (2.6.5)–(2.6.7) are equivalent as characterizing the same steady state. We wish to verify this equivalence and also relate the matrices $\mathbf{D}^{-1}\mathbf{B}$ and $\mathbf{A}$ for later use. It is first observed that Eqs.(1.2.9) and

$$\Delta H_j = \sum_{i=1}^{N} \nu_{ij} H_i \qquad j=1,\dots,R \tag{2.9.26}$$

can be combined to obtain $N+1-R$ independent linear relations among the rows of the matrix $\mathbf{D}^{-1}\mathbf{B}$, therefore this matrix has the number zero as an eigenvalue of multiplicity $N+1-R$. The remaining R eigenvalues of the matrix $\mathbf{D}^{-1}\mathbf{B}$ are the same as the eigenvalues of the matrix $\mathbf{A}$, $(A_{jk}=\partial \tilde{f}_j/\partial \xi_k)$. Indeed, if $\mathbf{v}=(v_1,\dots,v_{N+1})$ is an eigenvector of $\mathbf{D}^{-1}\mathbf{B}$ with eigenvalue τ, the vector $(w_1,\dots,w_R)$ defined by

$$D_i v_i = \sum_{j=1}^{R} \nu_{ij} w_j \qquad i=1,\dots,N, \tag{2.9.27}$$

$$k_c v_{N+1} = -\sum_{j=1}^{R} \Delta H_j w_j \tag{2.9.28}$$

is an eigenvector of the matrix $\mathbf{A}$, corresponding to the same eigenvalue τ. This can be verified in a straightforward way by using the relations

$$A_{jk} = \frac{\partial \tilde{f}_j}{\partial \xi_k} = \sum_{l=1}^{N+1} \frac{\partial f_j}{\partial u_l}\frac{\partial u_l}{\partial \xi_k} = \sum_{l=1}^{N} D_l^{-1} \nu_{lk} \frac{\partial f_j}{\partial c_l} - k_c^{-1} \Delta H_j \frac{\partial f_j}{\partial T}, \tag{2.9.29}$$

$$B_{ij} = \frac{\partial F_i}{\partial u_j} = \sum_{m=1}^{R} \nu_{im} \frac{\partial f_m}{\partial u_j} \qquad i=1,\dots,N, \tag{2.9.30}$$

$$B_{N+1,j} = \frac{\partial F_{N+1}}{\partial u_j} = -\sum_{m=1}^{R} \Delta H_m \frac{\partial f_m}{\partial u_j}. \tag{2.9.31}$$

Now if λ is an eigenvalue of Eqs.(2.9.24), (2.9.25) with eigenfunction χ then λ is also an eigenvalue of Eqs.(2.6.5)–(2.6.7) with an eigenfunction ϕ given by

$$D_i \chi_i = \sum_{j=1}^{R} \nu_{ij} \phi_j \qquad i=1,\dots,N, \tag{2.9.32}$$

$$k_c \chi_{N+1} = -\sum_{j=1}^{R} \Delta H_j \phi_j. \tag{2.9.33}$$

Obviously, the converse is also true, therefore the two problems (2.6.5)–(2.6.7) and (2.9.24), (2.9.25) have the same eigenvalues.

With this preparation, we will prove the first theorem.

Theorem 2.9.1. The steady state is stable if any of the following two conditions is satisfied

(i) The matrix $\mathbf{A}(\xi)$ defined by Eq.(2.6.4) is negative definite.

(ii) $r_0 \leqq (6/\alpha_M)^{\frac{1}{2}}$ (Eq.(2.7.9)).

Proof. By forming the scalar product of Eq. (2.9.22) with ψ, integrating and again integrating by parts there is obtained

$$\mu = \frac{-\int_0^{r_0} r^2 \left|\frac{d\psi}{dr}\right|^2 dr + \int_0^{r_0} r^2 (\mathbf{D}^{-1}\mathbf{B}\,\psi,\psi)\,dr}{\int_0^{r_0} r^2 (\mathbf{D}^{-1}\mathbf{P}\,\psi,\psi)\,dr}. \tag{2.9.34}$$

If $\mathbf{A}(\xi)$ is negative definite, $\mathbf{D}^{-1}\mathbf{B}(\mathbf{u})$ does not have eigenvalues with positive real parts and, as the denominator in Eq. (2.9.34) is positive, μ has a negative real part. Similarly, if $\alpha(\xi)$ is the largest positive eigenvalue of $\mathbf{A}^s$ (a matrix defined by Eq. (2.7.5)), $\alpha(\xi)$ is also the largest positive eigenvalue of $\mathbf{D}^{-1}\mathbf{B}$, therefore by using Poincaré's inequality, we show as in Section 2.7 that the numerator in Eq. (2.9.34) is less than

$$-\left(\frac{6}{r_0^2}-\alpha_M\right)\int_0^{r_0} r^2\,|\psi|^2\,dr \tag{2.9.35}$$

and Eq. (2.7.9) implies that μ has a negative real part. For the case of a single chemical reaction, Luss and Amundson [29] have used bifurcation point theory to prove the stronger result that a unique steady state is always stable. The extension of the bifurcation method to systems of reactions is an interesting subject for further research.

Index -1 Implies Instability

This property was obvious for the open uniform systems of Section 1.9, but is rather difficult to establish for distributed systems. We start by rewriting Eqs. (2.9.22), (2.9.23) and (2.9.24), (2.9.25) in operator form as

$$\psi - \mathbf{K}\,\psi = -\mu\,\mathbf{L}\,\psi, \tag{2.9.36}$$

$$\chi = \lambda\,\mathbf{K}\,\chi \tag{2.9.37}$$

where $\mathbf{K}, \mathbf{L}$ are the linear completely continuous operators defined by

$$\mathbf{K}\,\psi = \mathbf{D}^{-1}\int_0^{r_0} r'^2\,G(r,r')\,\mathbf{B}(\mathbf{u}(r'))\,\psi(r')\,dr', \tag{2.9.38}$$

$$\mathbf{L}\,\psi = \mathbf{D}^{-1}\,\mathbf{P}\int_0^{r_0} r'^2\,G(r,r')\,\psi(r')\,dr' \tag{2.9.39}$$

and $G(r,r')$ is the Green's function defined by Eq. (2.5.13). The operator $\mathbf{K}$ is in general nonselfadjoint, while the operator $\mathbf{L}$ is selfadjoint and positive.

In the following analysis, we shall use the real Hilbert space $\mathscr{B}_H$ of vector valued functions $\boldsymbol{\psi}=(\psi_1(r), \ldots, \psi_{N+1}(r))$ with inner product and norm

$$\langle\boldsymbol{\psi}, \boldsymbol{\phi}\rangle = \int_0^{r_0} r^2 (\boldsymbol{\psi}(r), \boldsymbol{\phi}(r))\, dr, \tag{2.9.40}$$

$$\|\boldsymbol{\psi}\| = \langle\boldsymbol{\psi}, \boldsymbol{\psi}\rangle^{\frac{1}{2}} \tag{2.9.41}$$

where

$$(\boldsymbol{\psi}, \boldsymbol{\phi}) = \sum_{i=1}^{N+1} \psi_i \phi_i. \tag{2.9.42}$$

We are now ready to proceed to our main subject:

Theorem 2.9.2. If a solution of Eqs. (2.4.1)–(2.4.4) has index -1, it is unstable.

Proof. An index -1 for a solution of Eqs. (2.4.1)–(2.4.4) implies that the vector field $\mathbf{I}-\mathbf{K}$ has rotation -1 on the sphere $\|\boldsymbol{\psi}\|=1$. Due to the complete continuity of the operator $\mathbf{K}$, a number $\tau>1$ can be chosen such that no eigenvalues λ of Eq. (2.9.37) lie in $[1, \tau]$. It follows from Theorem (A.5) that the vector field

$$\mathscr{H}_1 = \mathbf{I} - \tau\,\mathbf{K} \tag{2.9.43}$$

has also rotation -1 on the sphere $\|\boldsymbol{\psi}\|=1$. On the other hand, the vector field

$$\mathscr{H}(\mu) = \mathbf{I} + \frac{\mu\,\tau}{\tau-1}\,\mathbf{L} \tag{2.9.44}$$

has rotation $+1$ on the sphere $\|\boldsymbol{\psi}\|=1$ for each $\mu\in[0, \infty)$ because $\mathbf{L}$ has no negative eigenvalues. Since $\mathscr{H}_1$ and $\mathscr{H}(\mu)$ have different rotation, they are not homotopic, therefore given any $\mu\in[0, \infty)$, a number $q>0$ and a function $\boldsymbol{\psi}$, $\|\boldsymbol{\psi}\|=1$ can be found such that

$$\mathscr{H}_1\,\boldsymbol{\psi} = -q\,\mathscr{H}(\mu)\,\boldsymbol{\psi}. \tag{2.9.45}$$

Obviously, q and $\boldsymbol{\psi}$ are functions of μ:

$$q = q(\mu), \qquad \boldsymbol{\psi} = \boldsymbol{\psi}(\mu;\, r).$$

We wish to show the existence of a number $\underset{0}{\mu}>0$ such that $q(\underset{0}{\mu})=\tau-1$, for in this case, Eq. (2.9.45) reduces to Eq. (2.9.36) with $\mu=\underset{0}{\mu}$. To this end we shall show that

(i) $q(0)>\tau-1$.

(ii) $q(\mu)$ is continuous for each $\mu\in(0, \infty)$.

(iii) $\lim_{\mu\to\infty} q(\mu)=0$.

To prove item (i), we use the definition

$$\mathscr{H}_1\,\psi=-q(0)\,\mathscr{H}(0)\,\psi \tag{2.9.46}$$

and obtain

$$\psi=\frac{\tau}{1+q(0)}\,\mathbf{K}\,\psi \tag{2.9.47}$$

therefore

$$q(0)>\tau-1 \tag{2.9.48}$$

because Eq. (2.9.37) has no eigenvalues $\lambda\in[1,\tau]$. Item (ii) will be proved as a separate Theorem 2.9.3 below because it is lengthy and has independent interest.

To prove item (iii), we write Eq. (2.9.45) as

$$\frac{q\,\mu\,\tau}{\tau-1}\,\mathbf{L}\,\psi+(q+1)\,\psi=\tau\,\mathbf{K}\,\psi \tag{2.9.49}$$

and obtain

$$\left\|\frac{q\,\mu\,\tau}{\tau-1}\,\mathbf{L}\,\psi+(q+1)\,\psi\right\|=\tau\,\|\mathbf{K}\,\psi\|\,. \tag{2.9.50}$$

Consider a sequence $\mu_n\to\infty$ and the corresponding sequences $q(\mu_n)$, $\psi^{(n)}=\psi(\mu_n,r)$. If $\lim\limits_{\mu\to\infty}q(\mu)\neq 0$ then necessarily $\|\mathbf{L}\,\psi^{(n)}\|\to 0$ which implies $\|\mathbf{K}\,\psi^{(n)}\|\to 0$, as will be established in Lemma 2.9.1 at the end of the section. However, the quantity at the left of Eq. (2.9.50) is larger than $q+1$. This contradicts $\|\mathbf{K}\,\psi^{(n)}\|\to 0$ and hence the hypothesis $\lim\limits_{\mu\to\infty}q(\mu)\neq 0$ as well. Therefore, $\lim\limits_{\mu\to\infty}q(\mu)=0$.

Properties (i)–(iii) imply the existence of a number $\mu_0\in(0,\infty)$ such that $q(\mu_0)=\tau-1$ so that Eq. (2.9.45) yields

$$\psi-\mathbf{K}\,\psi=-\mu_0\,\mathbf{L}\,\psi\,. \tag{2.9.51}$$

Corollary 2.9.1. Among the $2m+1$ solutions of Eqs. (2.4.1)–(2.4.4), at least m are unstable.

Remark

If $\mathbf{B}$ is a symmetric matrix, then the following stronger version of Theorem 2.9.1 can be easily proved: If $\lambda\in(0,1)$ is an eigenvalue of Eq. (2.9.37), then Eq. (2.9.36) has a positive eigenvalue μ. Indeed, the quantity

$$\mu_0=\max_{\|\psi\|=1}\frac{\langle\mathbf{K}\,\psi,\psi\rangle-\langle\psi,\psi\rangle}{\langle\mathbf{L}\,\psi,\psi\rangle}$$

is an eigenvalue of Eq. (2.9.36). If $\lambda \in (0, 1)$ is an eigenvalue of Eq. (2.9.37) and χ the corresponding eigenfunction, then

$$\mu_0 \geqq \frac{\langle \mathbf{K}\,\chi, \chi\rangle - \langle \chi, \chi\rangle}{\langle \mathbf{L}\,\chi, \chi\rangle} = \frac{\left(\frac{1}{\lambda} - 1\right)\langle \chi, \chi\rangle}{\langle \mathbf{L}\,\chi, \chi\rangle} > 0 .$$

It appears very likely that this stronger result is also valid in the non-selfadjoint case, as with the open uniform systems of Section 1.9. One implication of the stronger result is that steady states with index $+1$ that have $2m'(m'>0)$ eigenvalues $\lambda \in (0, 1)$, are unstable.

Theorem 2.9.3. The function $q(\mu)$ defined by Eq. (2.9.45) is continuous in the interval $(0, \infty)$.

Proof. The proof is an application of KANTOROVICH'S implicit function theorem stated in the Appendix as Theorem A.6. The method used is similar to that of ROSENBLOOM [36]. We introduce a Hilbert space Y which is the cartesian product $Y = \mathscr{B}_H \times [-\infty, \infty]$, denote its elements by $\omega = (\psi, q)$, $\psi \in \mathscr{B}_H$, $q \in [-\infty, \infty]$, and define the inner product $\langle \omega, \omega'\rangle = \langle \psi, \psi'\rangle + q\,q'$ and the norm $\|\omega\| = \langle \omega, \omega\rangle^{\frac{1}{2}}$. The solution ψ of Eq. (2.9.45) will be normalized by

$$\langle \psi, \psi\rangle - 1 = 0 . \tag{2.9.52}$$

The *system of equations* (2.9.45), (2.9.52) can be denoted by

$$\mathscr{F}(\omega, \mu) = 0 \tag{2.9.53}$$

where the nonlinear operator $\mathscr{F}$ maps the product space $Y \times [-\infty, \infty]$ into the product space $\mathscr{B}_H \times [-\infty, \infty]$ and defines implicitly ω as a function of μ. Let now $\bar{\mu}, \bar{\omega} = (\bar{\psi}, \bar{q})$ satisfy

$$\mathscr{F}(\bar{\omega}, \bar{\mu}) = 0 . \tag{2.9.54}$$

We wish to show the local (in a neighborhood of $(\bar{\omega}, \bar{\mu})$) existence of a unique continuous function $\omega(\mu) = (\psi(\mu), q(\mu))$ such that $\omega(\bar{\mu}) = \bar{\omega}$ and $\mathscr{F}(\omega(\mu), \mu) = 0$. We have only to verify that the conditions of Theorem A.6 are satisfied.

The Fréchét derivative $\mathscr{F}_\omega$ at $\bar{\omega}, \bar{\mu}$ is computed by using the standard formula

$$\mathscr{F}_\omega(\bar{\omega}, \bar{\mu})\,\omega' = \lim_{\eta \to 0} \frac{\mathscr{F}(\bar{\omega} + \eta\,\omega', \bar{\mu}) - \mathscr{F}(\bar{\omega}, \bar{\mu})}{\eta} \tag{2.9.55}$$

where $\omega' = (\psi', q')$. Carrying out this differentiation on Eqs. (2.9.45), (2.9.52) we obtain $\mathscr{F}_\omega(\bar{\omega}, \bar{\mu})\,\omega' \in \mathscr{B}_H \times [-\infty, \infty]$ as

$$\mathscr{F}(\bar{\omega}, \bar{\mu})\,\omega' = \left(\mathscr{H}_1\,\psi' + \bar{q}\,\mathscr{H}(\bar{\mu})\,\psi' + q'\,\mathscr{H}(\bar{\mu})\,\bar{\psi},\ 2\langle \psi', \bar{\psi}\rangle\right). \tag{2.9.56}$$

We need to show that $\mathscr{F}_{\omega}^{-1}(\bar{\omega}, \bar{\mu})$ exists, in other words to show that the equations

$$\mathscr{H}_1 \psi' + \bar{q}\, \mathscr{H}(\bar{\mu})\, \psi' + q'\, \mathscr{H}(\mu)\, \bar{\psi} = \phi, \tag{2.9.57}$$

$$\langle \psi', \bar{\psi} \rangle = \alpha \tag{2.9.58}$$

can be solved in the unknown $\omega' = (\psi', q')$ for each $\phi \in \mathscr{B}_H, \alpha \in [-\infty, \infty]$. We first recall that

$$\mathscr{H}_1 \bar{\psi} + \bar{q}\, \mathscr{H}(\bar{\mu})\, \bar{\psi} = 0, \tag{2.9.59}$$

$$\mathscr{H}_1^* \bar{\psi}^* + \bar{q}\, \mathscr{H}(\bar{\mu})\, \bar{\psi}^* = 0 \tag{2.9.60}$$

where $\mathscr{H}_1^*$ is the adjoint of $\mathscr{H}_1$, i.e.,

$$\mathscr{H}_1^* \psi = \psi - \tau\, \mathbf{D}^{-1} \int_0^{r_0} r'^2\, \mathbf{B}^T(\mathbf{u}(r'))\, G(r, r')\, \psi(r')\, dr' \tag{2.9.61}$$

and $\mathbf{B}^T$ is the transpose of the matrix $\mathbf{B}$. Since $\mathscr{H}^{-1}(\mu)$ exists for all μ, we can rewrite Eqs. (2.9.59), (2.9.60) as

$$\mathscr{H}^{-1}(\bar{\mu})\, \mathscr{H}_1 \bar{\psi} + \bar{q}\, \bar{\psi} = 0, \tag{2.9.62}$$

$$(\mathscr{H}^{-1}(\bar{\mu})\, \mathscr{H}_1)^*\, \mathscr{H}(\bar{\mu})\, \bar{\psi}^* + \bar{q}\, \mathscr{H}(\bar{\mu})\, \bar{\psi}^* = 0. \tag{2.9.63}$$

Now the operator $\mathscr{H}^{-1}(\bar{\mu})\, \mathscr{H}_1$ is completely continuous as a product of a bounded and a completely continuous operator. It is well known from the spectral theory of completely continuous operators that Eqs. (2.9.59) – (2.9.63) imply

$$\langle \bar{\psi}, \bar{\psi}^* \rangle \neq 0, \tag{2.9.64}$$

$$\langle \bar{\psi}, \mathscr{H}(\bar{\mu})\, \bar{\psi}^* \rangle \neq 0. \tag{2.9.65}$$

Furthermore, the inhomogeneous equation (2.9.57) has a solution if and only if

$$\langle \bar{\psi}^*, \phi - q'\, \mathscr{H}(\bar{\mu})\, \bar{\psi} \rangle = 0. \tag{2.9.66}$$

This relation specifies q':

$$q' = \frac{\langle \bar{\psi}^*, \phi \rangle}{\langle \bar{\psi}^*, \mathscr{H}(\bar{\mu})\, \bar{\psi} \rangle} \tag{2.9.67}$$

since Eq. (2.9.65) insures the nonvanishing of the denominator. If ψ' is a solution of Eq. (2.9.57) then $\psi' + s\bar{\psi}$ is also a solution for any s. The number s is specified by Eq. (2.9.58)

$$s = \alpha - \langle \psi', \bar{\psi} \rangle. \tag{2.9.68}$$

We have now shown that $\mathscr{F}_{\omega}^{-1}$ exists as a linear operator, hence the conditions of Theorem A.6 are satisfied and the proof of Theorem 2.9.3 is complete.

Remark

We have implicitly assumed that μ is an eigenvalue of multiplicitly one. The case of higher multiplicitly can be treated in a very similar way.

Lemma 2.9.1. If $\psi^{(1)}, \psi^{(2)}, \ldots$ is a sequence of functions in $\mathscr{B}_H$ such that $\mathbf{L}\,\psi^{(n)} \to 0$, then $\mathbf{K}\,\psi^{(n)} \to 0$.

Proof. The proof will be restricted to scalar valued functions $\psi(r)$, for the case of vector valued functions does not present any additional difficulty except that of notation. Let

$$L\,\psi = \int_0^{r_0} r'^2\, G(r, r')\, \psi(r')\, dr', \tag{2.9.69}$$

$$K\,\psi = \int_0^{r_0} r'^2\, G(r, r')\, b(r')\, \psi(r')\, dr' \tag{2.9.70}$$

where $b(r)$ is continuous on $[0, r_0]$. The eigenfunctions of the operator L

$$\phi_m = \frac{1}{r} \sin \frac{m \pi r}{r_0} \qquad m = 1, 2, \ldots \tag{2.9.71}$$

form a complete orthogonal basis in $\mathscr{B}_H$. The sequence $\{\psi^{(n)}\}$ can be expanded in terms of the basis (2.9.71) as

$$\psi^{(n)} = \sum_{l=1}^{\infty} a_{nl}\, \phi_l \qquad n = 1, 2, \ldots. \tag{2.9.72}$$

By hypothesis, we have

$$\lim_{n\to\infty} L\,\psi^{(n)} = \lim_{n\to\infty} \frac{r_0^2}{\pi^2} \sum_{l=1}^{\infty} \frac{a_{nl}}{l^2}\, \phi_l = 0 \tag{2.9.73}$$

which implies

$$\lim_{n\to\infty} a_{nl} = 0 \qquad l = 1, 2, \ldots. \tag{2.9.74}$$

The next step is to evaluate $K\,\psi^{(n)}$:

$$K\,\psi^{(n)} = \sum_{l=1}^{\infty} a_{nl}\, K\,\phi_l, \tag{2.9.75}$$

$$K\,\phi_l = \int_0^{r_0} r'\, G(r, r')\, b(r') \sin \frac{l \pi r'}{r_0}\, dr'. \tag{2.9.76}$$

Since $b(r)$ is continuous, it can be expanded in the series

$$b(r) = \sum_{m=0}^{\infty} b_m \cos \frac{m \pi r}{r_0} \tag{2.9.77}$$

which when introduced in Eq. (2.9.76) gives

$$K\phi_l = \frac{1}{2}\sum_{m=0}^{\infty} b_m \int_0^{r_0} r'\, G(r,r')\left[\sin\frac{(l+m)\pi r'}{r_0} + \sin\frac{(l-m)\pi r'}{r_0}\right] dr'$$
$$= \frac{r_0^2}{2\pi^2}\left\{\sum_{m=0}^{\infty}\frac{b_m}{(m+l)^2}\frac{1}{r}\sin\frac{(l+m)\pi r}{r_0} + \sum_{\substack{m=0\\ m\neq l}}^{\infty}\frac{b_m}{(m-l)^2}\frac{1}{r}\sin\frac{(l-m)\pi r}{r_0}\right\}. \tag{2.9.78}$$

Therefore

$$K\psi^{(n)} = \frac{r_0^2}{2\pi^2}\left\{\sum_{m=0}^{\infty} b_m\left[\sum_{l=1}^{\infty}\frac{a_{nl}}{(m+l)^2}\right]\phi_m + \sum_{\substack{m=0\\ m\neq l}}^{\infty} b_m\left[\sum_{l=1}^{\infty}\frac{a_{nl}}{(m-l)^2}\,\mathrm{sign}(m-l)\right]\phi_m\right\}, \tag{2.9.79}$$

$$\|K\psi^{(n)}\| \leqq \frac{r_0^2}{2\pi^2}\left(\frac{r_0}{2}\right)^{\frac{1}{2}}[R_1+R_2] \tag{2.9.80}$$

where

$$R_1^2 = \sum_{m=0}^{\infty} b_m^2\left[\sum_{l=1}^{\infty}\frac{a_{nl}}{(m+l)^2}\right]^2, \tag{2.9.81}$$

$$R_2^2 = \sum_{\substack{m=0\\ m\neq l}}^{\infty} b_m^2\left[\sum_{l=1}^{\infty}\frac{a_{nl}}{(m-l)^2}\,\mathrm{sign}(m-l)\right]^2. \tag{2.9.82}$$

Now both the inner and outer series in Eqs. (2.9.81), (2.9.82) converge uniformly in n. Hence, the limit of these series as $n\to\infty$ can be taken inside the summations and by using Eq. (2.9.74) there is obtained

$$\lim_{n\to\infty} R_1 = 0, \qquad \lim_{n\to\infty} R_2 = 0. \tag{2.9.83}$$

Finally by taking the limit as $n\to\infty$ of Eq. (2.9.80) there is obtained

$$\lim_{n\to\infty} K\psi^{(n)} = 0. \tag{2.9.84}$$

Appendix

This appendix clarifies some matters of notation and states definitions and results of nonlinear functional analysis that were used in previous sections. The basic reference is KRASNOSEL'SKII [23].

Notation on Eigenvalue Problems

The term *characteristic* value is not used in this work. Thus, if $\mathbf{L}$ is a completely continuous linear operator, the values of the parameter λ for which the problem

$$\phi = \lambda \mathbf{L} \phi \tag{A.1}$$

has a solution are called *eigenvalues of equation* (A.1), instead of characteristic values of $\mathbf{L}$. Similarly, in the problems

$$\frac{1}{r^2} \frac{d}{dr} \left(r^2 \frac{d\phi}{dr} \right) = -\lambda' \mathbf{A} \phi, \tag{A.2}$$

$$\phi - \mathbf{K} \phi = -\mu \mathbf{L} \phi \tag{A.3}$$

where $\mathbf{K}, \mathbf{L}$ are completely continuous operators and $\mathbf{A}$ a matrix, λ', μ are called the eigenvalues of *equations* (A.2), (A.3) respectively.

Definition A.1. An operator $\mathbf{H}$ with domain and range in a Banach space $\mathscr{B}$ is called *completely continuous* if it is continuous and if for every bounded sequence $\xi^{(n)} \in \mathscr{B}$, the sequence $\{\mathbf{H}\, \xi^{(n)}\}$ contains a convergent subsequence. The term *compact* operator is not used in this work.

In the following we shall consider a closed, bounded and connected region $V \subset \mathscr{B}$, and shall denote by ∂V its boundary.

Definition A.2. The set of vectors $\xi - \mathbf{H}\,\xi$ is called the *vector field* associated with the operator $\mathbf{H}$ and is denoted by $\mathscr{H} = \mathbf{I} - \mathbf{H}$, where $\mathbf{I}$ is the identity operator. Often we can think of $\mathscr{H}$ as an operator with domain on a region V or its boundary ∂V. The vector field $\mathscr{H}$ is called *completely continuous* when the operator $\mathbf{H}$ is completely continuous.

Let $\mathscr{H}$ be a completely continuous vector field with no null vectors on ∂V, i.e. $\xi - \mathbf{H}\xi \neq 0$ for $\xi \in \partial V$. An integer γ, called the *rotation* of $\mathscr{H}$ on ∂V, can be uniquely defined [23]. The definition of γ involves considerable topological background and will not be given here in as much as we are interested only in the existence and some properties of γ.

Definition A.3. A point $\xi \in V$ such that $\mathcal{H}\xi = 0$, or equivalently $\xi = \mathbf{H}\,\xi$, is called a *null point of the vector field* $\mathcal{H}$ or a *fixed point of the operator* $\mathbf{H}$. The same point is also occasionally called *fixed point of the vector field* $\mathcal{H}$ in KRASNOSEL'SKII'S book. This last usage can cause confusion.

Definition A.4. The rotation of $\mathcal{H}$ on a sphere $\{\xi\colon \|\xi - \bar{\xi}\| = b\}$ centered at a null point $\bar{\xi}$ of $\mathcal{H}$ and containing no other null points of $\mathcal{H}$ is called the *index* of the point $\bar{\xi}$.

Theorem A.1 [23]. Let $\mathcal{H}$ be a completely continuous vector field having a finite number of null points in the interior of a region V and no null points on the boundary ∂V. Then the rotation of $\mathcal{H}$ on ∂V is equal to the sum of the indices of the null points. In particular, if the rotation differs from zero, there exists at least one null point of $\mathcal{H}$ (a fixed point of $\mathbf{H}$) in V.

The calculation of the rotation of a vector field is generally very difficult. In some cases, however, one can calculate the rotation by considering a simpler vector field. This is accomplished by using the very important concept of homotopy.

Definition A.5. Two completely continuous vector fields $\mathcal{H}_1, \mathcal{H}_2$, having no null points on the boundary ∂V of a region V, are called *homotopic* if there exists a completely continuous operator $\mathbf{H}(s)$, depending on a parameter $s \in [0, 1]$, and having the properties

$$\text{(i)}\quad \mathcal{H}_1 = \mathbf{I} - \mathbf{H}(0); \qquad \mathcal{H}_2 = \mathbf{I} - \mathbf{H}(1). \tag{A.4}$$

(ii) Given any $\varepsilon > 0$ there exists a $\delta > 0$ such that for any $\xi \in \partial V$ and $|s - s'| < \delta$:

$$\|\mathbf{H}(s)\,\xi - \mathbf{H}(s')\,\xi\| < \varepsilon. \tag{A.5}$$

(iii) The vector field $\mathcal{H}(s) = \mathbf{I} - \mathbf{H}(s)$ has no null points on ∂V for $s \in [0, 1]$. The following theorem is the basis of calculating the rotation of many fields:

Theorem A.2 [23]. Two vector fields have the same rotation if and only if they are homotopic.

Theorem A.3 [23]. The rotation of the vector field $\mathcal{H} = \mathbf{I}$ is zero if $0 \notin \mathrm{Int}(V)$ and is $+1$ if $0 \in \mathrm{Int}(V)$ (it is not defined if $0 \in \partial V$).

Theorem A.4 [23]. Consider the Euclidean space E_{N+1}. The rotations γ^+, γ^- of the vector fields $\mathcal{H}$, $-\mathcal{H}$ on the N-dimensional surface ∂V of a region V are related by $\gamma^- = (-1)^{N+1}\,\gamma^+$. In particular, the rotation of the field $-\mathbf{I}$ on a sphere surrounding the origin is $(-1)^{N+1}$.

It is often necessary to calculate the index of a fixed point. For a linear vector field, the index can be obtained in terms of the eigenvalues of the linear operator. A nonlinear vector field can be shown to be locally homotopic with the linear field obtained by linearization. Before stating the basic result we have to define precisely the linearization process.

Definition A.6. A continuous operator $\mathbf{H}$ is called differentiable at a point $\xi \in \mathscr{B}$ if the increment $\mathbf{H}(\xi+\phi)-\mathbf{H}\,\xi$ can be expressed as

$$\mathbf{H}(\xi+\phi)-\mathbf{H}\,\xi=\mathbf{L}\,\phi+\mathbf{r}(\xi,\phi) \tag{A.6}$$

where $\mathbf{L}$ is a linear operator and

$$\lim_{\|\phi\|\to 0} \frac{\|\mathbf{r}(\xi,\phi)\|}{\|\phi\|}=0. \tag{A.7}$$

The operator $\mathbf{L}$, which depends on ξ, is called the Fréchét derivative of $\mathbf{H}$ at the point ξ. In [23] it is proved that the Fréchét derivative of a completely continuous operator is completely continuous. When the Fréchét derivative exists it can be computed from the formula ([20], p. 657)

$$\mathbf{L}\,\phi=\lim_{s\to 0} \frac{\mathbf{H}(\xi+s\,\phi)-\mathbf{H}\,\xi}{s}. \tag{A.8}$$

Theorem A.5 [23], p. 136. Let ξ be a fixed point of a completely continuous operator $\mathbf{H}$ and let $\mathbf{L}$ be the Fréchét derivative of $\mathbf{H}$ at ξ. Furthermore suppose that $\lambda=1$ is not an eigenvalue of the problem

$$\phi=\lambda\,\mathbf{L}\,\phi. \tag{A.9}$$

Then ξ is an isolated fixed point of $\mathbf{H}$ with index $(-1)^{\beta}$, where β is the sum of multiplicities of the eigenvalues of Eq. (A.9) (characteristic values of $\mathbf{L}$) that lie in the interval [0, 1].

The proof of this theorem is obtained by linearizing the equation $\xi=\mathbf{H}\,\xi$ to $\phi=\mathbf{L}\,\phi$ in a sufficiently small neighborhood of the fixed point ξ. The vector fields $\mathbf{I}-\mathbf{H}, \mathbf{I}-\mathbf{L}$ are then shown to be homotopic and therefore to have the same index. Now, the rotation of any completely continuous linear vector field $\mathbf{I}-\mathbf{L}$ on a small sphere around ξ has been calculated in [23] as $(-1)^{\beta}$, therefore this must also be the rotation of the field $\mathbf{I}-\mathbf{H}$. When $\lambda=1$ is an eigenvalue of Eq. (A.9), then linearization of $\mathbf{H}$ is not legitimate because ξ is a "double root" of the equation $\xi=\mathbf{H}\,\xi$. A small change in the operator $\mathbf{H}$ will either separate the double root to two neighboring "simple" roots or will annihilate it. More specifically, we may consider the equation $\xi-\mathbf{H}(\alpha)\,\xi=0$ (α is a parameter) as defining implicitly $\xi(\alpha)$. Then $\lambda=1$ is the exceptional case in which the implicit function theorem fails to apply. In the case of $\lambda=1$ Eq. (A.9) can be written as

$$\phi=\mathbf{L}(\xi(\alpha))\,\phi \tag{A.10}$$

and considered as an equation in α. This equation is called the bifurcation equation and is treated among others by KRASNOSEL'SKII [23] and NIRENBERG [33].

The following remark clarifies a minor point relative to the applicability of Theorem A.5. In Sections 1.4, 2.6 we have extended the definition of f_j outside their natural domain $\tilde{\Gamma}(u_0)$. The extension is continuous, but has discontinuous derivatives on $\partial\tilde{\Gamma}$. However, the one sided derivatives exist, therefore the operator $\mathbf{H}$ defined by Eq. (2.5.16) has one sided Fréchét derivatives on $\partial V(u_0)$ (a region defined by Eq. (2.5.18)). The one sided derivative taken from the interior is defined by

$$\mathbf{H}(\xi+\phi)-\mathbf{H}\,\xi=\mathbf{L}\,\phi+\mathbf{r}(\xi,\phi)$$

where $\mathbf{r}$ satisfies Eq. (A.7) for all $\xi\in V(u_0)$, $\xi+\phi\in V(u_0)$. It can be shown easily that inspite the discontinuity of $\mathbf{L}$ on $\partial V(u_0)$, Theorem A.5 applies to fixed points $\xi\in\partial V(u_0)$ with $\mathbf{L}$ being the interior one sided derivative of $\mathbf{H}$. This extension of Theorem A.5 is based on the fact that $\mathbf{H}$ has no fixed points outside of $V(u_0)$.

Theorem A.6, KANTOROVICH [20], p. 686. Let $\mathscr{F}$ be an operator defined on a neighborhood N of the point $(\bar{\omega},\bar{\mu})\in Y\times X$, mapping N into Z, and continuous at $(\bar{\omega},\bar{\mu})$, (Y, X, Z are Banach spaces). If

(i) $\mathscr{F}(\bar{\omega},\bar{\mu})=0$,

(ii) the Fréchét derivative $\mathscr{F}_\omega(\omega,\mu)$ exists in N and is continuous at $\bar{\omega},\bar{\mu}$,

(iii) the operator $\mathscr{F}_\omega(\bar{\omega},\bar{\mu})$, mapping Y into Z, has a linear inverse,

then a unique continuous function $\omega(\mu)$, satisfying $\mathscr{F}(\omega(\mu),\mu)=0$, $\omega(\bar{\mu})=\bar{\omega}$, exists in some neighborhood of $\bar{\mu}$.

Notation

A_j, $\tilde{A}_j$	the affinity of the jth reaction, Eq. (1.5.1)
$\mathbf{A}$	a matrix, Eqs. (1.7.5), (2.6.4)
A_{jk}	$A_{jk}=(\partial\tilde{f}_j/\partial\xi_k)$, Eqs. (1.7.5), (2.6.4)
$\mathscr{A}_j$	the jth atomic species
$\mathbf{B}$, B_{ij}	a matrix and its elements defined by Eqs. (2.9.30), (2.9.31)
$\mathscr{B}$	a Banach space
$\mathscr{B}_c$	the Banach space of continuous functions
$\mathscr{B}_H$	a Hilbert space, Eqs. (2.9.40), (2.9.41)
c_i	the concentration of the ith species
c_{i0}	initial (Chapter 1) or surface (Chapter 2) concentration of the ith species
c_{ia}	concentration of the ith species in the ambient fluid
c_i^*	equilibrium concentration of the ith species
c_{if}	concentration of the ith species in the input stream
c_{im}, c_{iM}	quantities defined by Eq. (1.2.17) or Eq. (2.8.1)
c_{vi}, c_{pi}	heat capacities of the ith species under constant volume or pressure
c_p	heat capacity of a catalyst pellet
C_p	$C_p=\sum c_{pi}c_i$
$\mathbf{D}$	a matrix defined by Eq. (2.9.20)
D_i	the diffusivity (effective) of the ith species
$d(\mathbf{u})$	the distance from the point $\mathbf{u}$ to the surface ∂E_{N+1}^+
E_{N+1}, E_{N+1}^+	the $N+1$-dimensional Euclidean space and its positive orthant
E_R	the R-dimensional Euclidean space
E, E_1, E_2	activation energies
F_i	the rate of production of the ith species by chemical reactions, Eq. (1.1.4)
F_{iM}	a quantity defined by Eq. (2.8.4)
$\mathscr{F}$	an operator introduced in Section 2.9, Eq. (2.9.53)
$\mathscr{F}_\omega$	the Fréchét derivative of $\mathscr{F}$
$\tilde{\mathbf{f}}$	$\tilde{\mathbf{f}}=(\tilde{f}_1,\ldots,\tilde{f}_R)$, the reaction rate vector
$f_j(\mathbf{u})$	the rate of the jth reaction
$\tilde{f}_j(\xi)$	the rate of the jth reaction, Eq. (1.2.25)
f_{jM}	a quantity defined by Eq. (2.8.4)
f_j'	the rate of a surface reaction
$G(r, r')$	a GREEN's function defined by Eq. (2.5.13)
$\mathbf{H}$	a nonlinear operator
H_i	the enthalpy of the ith species

h, h_i, $\tilde{h}_j$	Thiele moduli defined by Eqs. (2.3.14), (2.8.17), (2.8.19)
$\mathscr{H}$	$\mathscr{H} = \mathbf{I} - \mathbf{H}$, the vector field corresponding to the operator $\mathbf{H}$
$\mathscr{H}(\mu)$, $\mathscr{H}_1$	vector fields introduced in Section 2.9
$\mathbf{I}$	the identity operator
J	a function defined by Eq. (2.3.19)
$\mathbf{K}$	an operator defined by Eq. (2.9.38)
$K_j(T)$	equilibrium constant, Eq. (1.6.12)
k_i	a reaction rate constant
k_c	the thermal conductivity of a catalyst pellet
$\mathbf{L}$	the Fréchét derivative of the nonlinear operator $\mathbf{H}$, also in Section 2.9, an operator defined by Eq. (2.9.39)
l_i	the film transfer coefficient of the ith species
M	the number of atomic species
M_i	the molecular weight of the ith species
$\mathscr{M}_i$	a symbol for the ith chemical species
N	the number of chemical species
$\mathbf{P}$	a matrix defined by Eq. (2.9.21)
p	the pressure
p_0	a reference pressure
Q	total rate of heat exchange for a stirred-tank reactor
$q(\mu)$	a function introduced by Eq. (2.9.45) and used in Section 2.9 and the Appendix
R	the number of chemical reactions
R_g	the universal gas constant
R_β	the rank of the matrix $\boldsymbol{\beta}$, Eq. (1.1.9)
r	the radial coordinate
r_0	the radius of a catalyst pellet
r_p	the radius of a catalyst pore
S, $\tilde{S}$	in Sections 1.5, 1.6 indicate the entropy
S, S_1	spheres in a finite or infinite-dimensional space
S_i	the entropy of the ith species, Section 1.6
S_p	surface area per unit of catalyst mass
T	the temperature
T_0	initial (Chapter 1) or surface (Chapter 2) temperature
T_a	temperature of the ambient fluid
T^*	equilibrium temperature
T_f	temperature of the input stream
T_m, T_M	quantities defined by Eq. (1.2.18) or Eq. (2.8.2)
t	the time
U, U_0	the internal energy
$\mathbf{u}$	$\mathbf{u} = (c_1, \ldots, c_N, T)$, the state (vector)
$\mathbf{u}_0$	$\mathbf{u}_0 = (c_{10}, \ldots, c_{N0}, T_0)$, the initial (Chapter 1) or surface (Chapter 2) state

$\mathbf{u}_a$	$\mathbf{u}_a=(c_{1a},\ldots,c_{Na},T_a)$, the state of the ambient fluid
$\mathbf{u}^*$	$\mathbf{u}^*=(c_1^*,\ldots,c_N^*,T^*)$, the equilibrium state
$\mathbf{u}_f$	$\mathbf{u}_f=(c_{1f},\ldots,c_{Nf},T_f)$, the state of the input stream
$\mathbf{u}_s$	the steady state
V	the volume of a stirred-tank reactor (Section 1.8), also a region in $\mathscr{B}_c$
∂V	the boundary of the region V
$V(u_0)$	a region defined by Eq.(1.3.6) or Eq.(2.5.18)
$V_1(u_0)$	a region defined by Eq.(1.3.8)
W	a function defined by Eq.(1.2.8)
w	the volumetric flow rate in a stirred-tank reactor
x	the spatial coordinate in a planar region
x_0	the half thickness of a planar region
$\alpha(\xi)$, α_M	quantities defined below, Eq.(2.7.5)
β_{il}	the number of atoms $\mathscr{A}_l$ in the species $\mathscr{M}_i$
β	the sum of multiplicities of eigenvalues in the interval [0, 1]
$\Gamma(u_0)$	the invariant manifold including $\mathbf{u}_0$
$\tilde{\Gamma}(u_0)$	the image of $\Gamma(u_0)$ in the ξ space
$\tilde{\Gamma}^+(u_0)$	the positive section of $\tilde{\Gamma}(u_0)$
γ	the rotation of the vector field $\mathscr{H}$
γ_{il}	coefficients defined by Eq.(1.2.9)
ΔH_j	the heat of the jth reaction, Eq.(1.8.3)
ΔF_j^0	a quantity defined by Eq.(1.6.10)
Δv_j	a quantity defined by Eq.(1.6.14)
E	effectiveness factor for a single reaction
$\mathsf{E}^{(1)}$, $\tilde{\mathsf{E}}^{(3)}$	effectiveness factors for planar and spherical regions
$\mathsf{E}_i^{(3)}$, $\mathsf{E}_j^{(3)}$	effectiveness factors for the ith species and the jth reaction in spherical regions
ε_p	void fraction of a catalyst pellet
ζ	dimensionless position coordinate, Eq.(2.3.10)
η	dimensionless concentration, Eq.(2.3.10)
θ	$\theta=V/w$, holding time
λ, λ'	eigenvalues
μ	eigenvalue, a parameter in Section 2.9
μ_i	the chemical potential of the ith species
$\mathbf{v}$	the stoichiometric matrix
v_i, v_{ij}	stoichiometric coefficients introduced by Eqs.(1.1.1), (1.1.3)
ξ_j	the extent of the jth reaction
$\boldsymbol{\xi}$	$\boldsymbol{\xi}=(\xi_1,\ldots,\xi_R)$, the state described in terms of the extents
$\boldsymbol{\xi}_s$	the steady state
$\boldsymbol{\xi}^*$	the equilibrium state
ξ_{jm}, ξ_{jM}	quantities defined by Eq.(2.8.3)
ρ_p	apparent density of a catalyst pellet
σ	a positive definite form defined by Eq.(2.8.21), the entropy production

τ	a parameter introduced in Th.(2.9.2)
$\phi(\eta)$	a dimensionless reaction rate, Eq.(2.3.10)
$\boldsymbol{\phi}$	$\boldsymbol{\phi}=(\phi_1, \ldots, \phi_R)$, an eigenvector or eigenfunction
$\Phi(\eta, \eta')$	a function defined by Eq.(2.3.18)
Φ_1, Φ_2	quantities defined by Eq.(2.8.59), (2.8.60)
$\boldsymbol{\chi}$	an eigenfunction
$\boldsymbol{\psi}$	an eigenfunction
$\bar{\boldsymbol{\psi}}$	a solution of Eq.(2.9.59)
$\bar{\boldsymbol{\psi}}^*$	the solution of Eq.(2.9.60)
Ω, Ω'	the thermodynamic and kinetic equilibrium manifolds
$\mathbf{X}(\boldsymbol{\psi}, q)$	an element of a product space, introduced in Th.(2.9.3)

References

1. AMDUR, I., and G. G. HAMMES: Chemical kinetics: principles and selected topics. New York: McGraw-Hill Book Co. 1966.
2. AMUNDSON, N. R., and L. R. RAYMOND: Stability in distributed parameter systems. A.I.Ch.E. Journal **11**, 339 (1965).
3. ARIS, R.: Introduction to the analysis of chemical reactors. New Jersey: Prentice-Hall, Inc. 1965.
4. – Prolegomena to the rational analysis of systems of chemical reactions. Arch. Rational Mech. Anal. **19**, 81 (1965).
5. – Prolegomena to the rational analysis of systems of chemical reactions. II. Some addenda. Arch. Rational Mech. Anal. **27**, 356 (1968).
6. –, and N. R. ADMUNDSON: An analysis of chemical reactor stability and control. I. The possibility of local control, with perfect or imperfect control mechanisms. Chem. Eng. Sci. **7**, 121 (1958).
7. – – An analysis of chemical reactor stability and control. III. The principles of programming reactor calculations. Some extensions. Chem. Eng. Sci. **7**, 148 (1958).
8. –, and R. H. S. MAH: Independence of chemical reactions. Ind. Eng. Chem. Fundamentals **2**, 90 (1963).
9. BOWEN, J. R., A. ACRIVOS, and A. K. OPPENHEIM: Singular perturbation refinement to quasi-steady state approximation in chemical kinetics. Chem. Eng. Sci. **18**, 177 (1963).
10. BOWEN, R. M.: On the stoicheiometry of chemically reacting materials. Arch. Rational Mech. Anal. (In press.)
11. – On the thermochemistry of reacting materials. Report SC-DC-67-2389.
12. BOX, G. E. P.: Fitting empirical data. Ann. N.Y. Acad. Sci. **86**, 792 (1960).
13. DE GROOT, S. R., and P. MAZUR: Non-equilibrium thermodynamics. Amsterdam: North-Holland Publ. Co. 1962.
14. DENBIGH, K.: The principles of chemical equilibrium. New York: Cambridge University Press 1963.
15. FROST, A. A., and R. G. PEARSON: Kinetics and mechanism, 2nd ed. New York: Wiley & Sons 1961.
16. GAVALAS, G. R.: On the steady states of distributed parameter systems with chemical reactions, heat, and mass transfer. Chem. Eng. Sci. **21**, 477 (1966).
17. – The behavior of distributed systems with complex reactions and diffusion in the regime of transport limitation. Chem. Eng. Sci. **22**, 997 (1967).
18. HEINEKEN, F. G., H. M. TSUCHIYA, and R. ARIS: On the mathematical status of the pseudo-steady state hypothesis of biochemical kinetics. Math. Biosci **1**, 95 (1967).
19. HORIUTI, J.: Significance and experimental determination of stoichiometric number. J. Catalysis **1**, 199 (1962).
20. KANTOROVICH, L. V., and G. P. AKILOV: Functional analysis in normed spaces. New York: Pergamon Press 1964.

21. KATO, T.: Perturbation theory for linear operators. New York: Springer 1966.
22. KITTRELL, J. R., R. MEZAKI, and C. C. WATSON: Estimation of parameters for nonlinear least squares analysis. Ind. Eng. Chem. Fundamentals **57**, 12, 18 (1965).
23. KRASNOSEL'SKII, M. A.: Topological methods in the theory of nonlinear integral equations. New York: Pergamon Press 1964.
24. KUO, J. C. W., and N. R. AMUNDSON: Catalytic particle stability studies – I. Lumped resistance model. Chem. Eng. Sci. **22**, 49 (1967).
25. – – Catalytic particle stability studies – II. Lumped thermal resistance model. Chem. Eng. Sci. **22**, 443 (1967).
26. – – Catalytic particle stability studies – III. Complex distributed resistance model. Chem. Eng. Sci. **22**, 1185 (1967).
27. LAIDLER, K. J.: Chemical kinetics, 2nd ed. New York: McGraw-Hill Book Co. 1965.
28. LEE, E. S.: Quasilinearization and estimation of parameters in differential equations. Ind. Eng. Chem. Fundamentals **7**, 152 (1968).
29. LUSS, D., and N. R. AMUNDSON: Some general observations on tubular reactor stability. Can. J. Chem. Eng. **45**, 341 (1967).
30. – – Uniqueness of the steady state solutions for chemical reaction occurring in a catalyst particle or in a tubular reactor with axial diffusion. Chem. Eng. Sci. **22**, 253 (1967).
31. – Sufficient conditions for uniqueness of the steady state solutions in distributed parameter systems. Chem. Eng. Sci. (In press.)
32. LUUS, R., and L. LAPIDUS: An averaging technique for stability analysis. Chem. Eng. Sci. **21**, 159 (1966).
33. NIRENBERG, L.: Functional analysis. Lecture Notes. New York University 1960/61.
34. PETERSEN, E. E.: Chemical reaction analysis. New Jersey: Prentice-Hall, Inc. 1965.
35. PRIGOGINE, I., and R. DEFAY: Chemical thermodynamics. New York: Longmans Green & Co. 1954.
36. ROSENBLOOM, P.: Perturbation of linear operators in Banach spaces. Arch. Math. **6**, 89 (1955).
37. SELLERS, P. H.: Algebraic complexes applied to chemistry. Proc. Nat. Acad. Sci. U.S. **55**, 693 (1966).
38. – Algebraic complexes which characterize chemical networks. Soc. Ind. Appl. Math. J. **15**, 13 (1967).
39. VAN RYSSELBERGHE, P.: Reaction rates and affinities. J. Chem. Phys. **29**, 640 (1958).
40. WEI, J.: Axiomatic treatment of chemical reaction systems. J. Chem. Phys. **36**, 1578 (1962).
41. – Intraparticle diffusion effects in complex systems of first order reactions. I. The effects in single particles. J. Catalysis **1**, 526 (1962).
42. – The stability of a reaction with intra-particle diffusion of mass and heat: The Liapunov methods in a metric function space. Chem. Eng. Sci. **20**, 729 (1965).
43. –, and C. D. PRATER: The structure and analysis of complex reaction systems. Advances in Catalysis **13**, 203 (1962).

Subject Index

Springer Tracts in Natural Philosophy

1 **Gundersen: Linearized Analysis of One-Dimensional Magnetohydrodynamic Flows**
With 10 figures. X, 119 pages. 1964. Cloth DM 22,—; US $ 5.50

2 **Walter: Differential- und Integral-Ungleichungen und ihre Anwendung**
bei Abschätzungs- und Eindeutigkeitsproblemen
Mit 18 Abbildungen. XIV, 269 Seiten. 1964. Gebunden DM 59,—; US $ 14.75

3 **Gaier: Konstruktive Methoden der konformen Abbildung**
Mit 20 Abbildungen und 28 Tabellen. XIV, 294 Seiten. 1964
Gebunden DM 68,—; US $ 17.00

4 **Meinardus: Approximation von Funktionen und ihre numerische Behandlung**
Mit 21 Abbildungen. VIII, 180 Seiten. 1964. Gebunden DM 49,—; US $ 12.25

5 **Coleman, Markovitz, Noll: Viscometric Flows of Non-Newtonian Fluids**
Theory and Experiment
With 37 figures. XII, 130 pages. 1966. Cloth DM 22,—; US $ 5.50

6 **Eckhaus: Studies in Non-Linear Stability Theory**
With 12 figures. VIII, 117 pages. 1965. Cloth DM 22,—; US $ 5.50

7 **Leimanis: The General Problem of the Motion of Coupled Rigid Bodies about a Fixed Point**
With 66 figures. XVI, 337 pages. 1965. Cloth DM 48,—; US $ 12.00

8 **Roseau: Vibrations non linéaires et théorie de la stabilité**
Avec 7 figures. XII, 254 pages. 1966. Pleine toile DM 39,—; US $ 9.75

9 **Brown: Magnetoelastic Interactions**
With 13 figures. VIII, 155 pages. 1966. Cloth DM 38,—; US $ 9.50

10 **Bunge: Foundations of Physics**
With 5 figures. XII, 311 pages. 1967. Cloth DM 56,—; US $ 14.00

11 **Lavrentiev: Some Improperly Posed Problems of Mathematical Physics**
With 1 figure. VIII, 72 pages. 1967. Cloth DM 20,—; US $ 5.00

12 **Kronmüller: Nachwirkung in Ferromagnetika**
Mit 92 Abbildungen. XIV, 329 Seiten. 1968. Gebunden DM 62,—; US $ 15.50

13 **Meinardus: Approximation of Functions: Theory and Numerical Methods**
With 21 figures. VIII, 198 pages. 1967. Cloth DM 54,—; US $ 13.50

14 **Bell: The Physics of Large Deformation of Crystalline Solids**
With 166 figures. X, 253 pages. 1968. Cloth DM 48,—; US $ 12.00

15 **Buchholz: The Confluence Hypergeometric Function with special emphases on its applications.** In preparation

16 **Slepian: Mathematical Foundations of Network Analysis**
Approx. 216 pages. 1968. Cloth DM 44,—; US $ 11.00

Universitätsdruckerei H. Stürtz AG Würzburg